Math Mammoth
Grade 1 Answer Keys

for the complete curriculum
(Light Blue Series)

Includes answer keys to:

- Worktext part A
- Worktext part B
- Tests
- Cumulative Reviews

By Maria Miller

Contents

Math Mammoth Grade 1-A
Answer Key

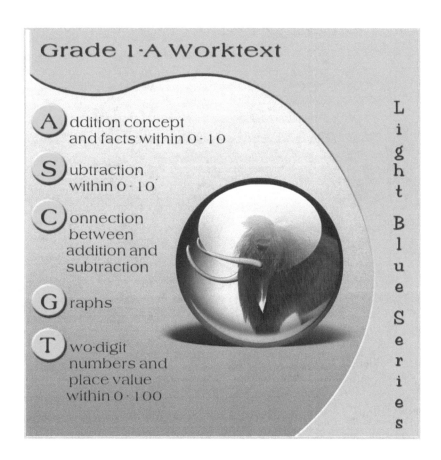

By Maria Miller

Contents

Chapter 0: Kindergarten Review

Equal Amounts; Same and Different, p. 7

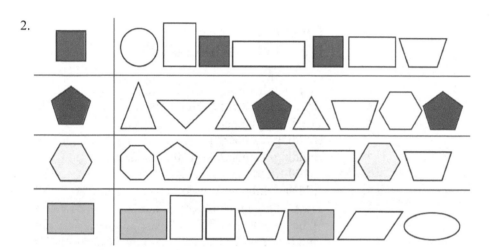

Writing Numbers, p. 8

1. Check the student's work.

2. a. 4   b. 2   c. 0   d. 3   e. 1   f. 2

3. Check the student's work.

4. a. 6   b. 5   c. 3   d. 7   e. 8   f. 9

Counting, p. 10

1. a. 3   b. 2   c. 5   d. 4   e. 6   f. 5   g. 9   h. 8

2. a. 7, 6   7 is more   b. 3, 4   4 is more   c. 12, 10   12 is more   d. 6, 7   7 is more

3. a. 4   b. 9   c. 8

4. a. 4 is more; the total is 6.   b. 6 is more; the total is 11.   c. 6 is more; the total is 10.
   d. 10 is more; the total is 19.   e. 8 is more; the total is 14.   f. 7 is more; the total is 13.

1. a. 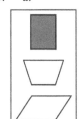 b.  c.

2. a.

   b.

   c.

   d.

   e.

3. a.  b. 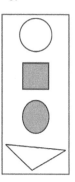 c.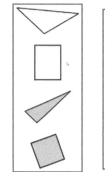

4. a.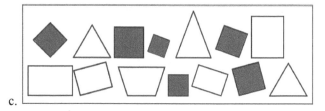

   b.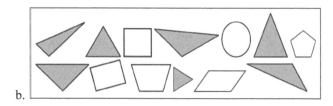

   c.

Patterns, p. 14

1. a. A white triangle.   b. A white square.   c. A white square.   d. A purple square.
   e. A yellow square.   f. A red circle.   g. A blue circle.

2. 5, 6, 7, 8, 9, 10
   5, 4, 3, 2, 1, 0

Chapter 1: Addition Within 0-10

Two Groups and a Total, p. 19

1.

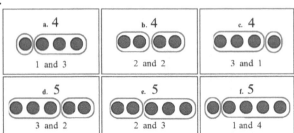

2. a. 3   b. 2   c. 1
   d. 1   e. 2   f. 3
   g. 4   h. 0   i. 5

3. The answers will vary.
   Please check the student's work.

4. a. 1 and 2   b. 4 and 2   c. 2 and 3
   d. 3 and 1   e. 3 and 3   f. 1 and 4

5. a. $2 + 1 = 3$ b. $3 + 1 = 4$ c. $2 + 2 = 4$
   d. $2 + 3 = 5$ e. $1 + 3 = 4$ f. $1 + 1 = 2$
   g. $3 + 2 = 5$ h. $4 + 1 = 5$ i. $1 + 2 = 3$

6. a. 4   b. 4   c. 6   d. 5

Learn the Symbols + and =, p. 22

1. a. 4   b. $1 + 2 = 3$
   c. $3 + 2 = 5$ d. $1 + 4 = 5$
   e. $2 + 3 = 5$ f. $1 + 1 = 2$
   g. $2 + 2 = 4$ h. $3 + 2 = 5$
   i. $3 + 1 = 4$ j. $2 + 2 = 4$

2. a. $1 + 3 = 4$ b. $3 + 2 = 5$
   c. $2 + 3 = 5$ d. $2 + 1 = 3$

3. a. $2 + 0 = 2$ b. $3 + 0 = 3$ c. $4 + 0 = 4$
   d. $0 + 2 = 2$ e. $0 + 5 = 5$ f. $1 + 0 = 1$
   g. $0 + 3 = 3$ h. $0 + 0 = 0$

4. a. $1 + 3 = 4$ b. $2 + 2 = 4$ c. $4 + 1 = 5$
   d. $2 + 0 = 2$ e. $3 + 2 = 5$ f. $0 + 1 = 1$
   g. $2 + 1 = 3$ h. $3 + 0 = 3$ i. $1 + 1 = 2$
   j. $2 + 3 = 5$

Addition Practice 1, p. 25

1. a. $2 + 1 = 3$ b. $3 + 2 = 5$ c. $1 + 2 = 3$
   d. $4 + 1 = 5$ e. $2 + 3 = 5$ f. $0 + 4 = 4$
   g. $2 + 2 = 4$ h. $1 + 0 = 1$ i. $3 + 1 = 4$

2. a. $2 + 2 = 4$ b. $1 + 3 = 4$ c. $0 + 5 = 5$
   d. $4 + 1 = 5$ e. $2 + 3 = 5$ f. $1 + 3 = 4$

3. a. $1 + 2 = 3$ b. $3 + 0 = 3$ c. $2 + 2 = 4$
   d. $2 + 3 = 5$ e. $1 + 4 = 5$ f. $0 + 5 = 5$
   g. $3 + 2 = 5$ h. $2 + 1 = 3$ i. $4 + 1 = 5$

4. b. $1 + 2 = 3, 2 + 1 = 3$ c. $3 + 1 = 4, 1 + 3 = 4$
   d. $1 + 4 = 5, 4 + 1 = 5$ e. $0 + 2 = 2, 2 + 0 = 2$
   f. $5 + 0 = 5, 0 + 5 = 5$

Which Is More?, p. 27

2. a. 3   b. 5   c. 5
   d. 6; 6 is greater than 2.
   e. 4; 4 is greater than 1.
   f. 4; 2 is less than 4.

3. a. 6   b. 4   c. 5   d. 4
   e. 2   f. 3   g. 5   h. 4

4. a. $1 < 4$ b. $2 < 5$ c. $6 > 3$
   d. $3 < 4$ e. $5 > 1$ f. $2 < 3$

5. a. $1 < 4$ b. $4 > 3$ c. $2 < 5$ d. $0 < 4$

6. a. $1 < 4$ b. $4 < 5$ c. $2 < 4$ d. $5 > 3$
   e. $1 < 2$ f. $3 > 1$ g. $5 > 4$ h. $4 < 6$
   i. $3 < 5$ j. $1 > 0$ k. $2 < 5$ l. $0 < 2$

Missing Items, p. 29

1. a. 2   b. 1   c. 1   d. 2
   e. 3   f. 1   g. 0   h. 3   i. 2

2. a. 2   b. 0   c. 4   d. 2   e. 3   f. 1
   g. 0   h. 2   i. 3   j. 4   k. 0   l. 1

3. a. 2   b. 0   c. 1   d. 3   e. 2   f. 1

4. a. 4   b. 0   c. 3   d. 0   e. 2   f. 1

5. a. 1   b. 1   c. 2   d. 1   e. 1   f. 2   g. 4   h. 3   i. 0

6. a. 3   b. 2   c. 2   d. 0   e. 2   f. 2   g. 1   h. 2   i. 4

7. a. 2, 3   b. 4, 4   c. 5, 4   d. 5, 5   e. 5, 3   f. 5, 5

8. a. 3, 2, 1, 0   b. 4, 3, 2, 1, 0   c. 5, 4, 3, 2, 1, 0

Sums with 5, p. 34

1.

$0 + 5 = 5$	$5 + 0 = 5$
$1 + 4 = 5$	$4 + 1 = 5$
$2 + 3 = 5$	$3 + 2 = 5$

2.

a. $4 + 1 = 5$	b. $2 + 3 = 5$	c. $1 + 1 = 2$
$2 + 2 = 4$	$1 + 3 = 4$	$0 + 5 = 5$
$3 + 2 = 5$	$1 + 4 = 5$	$1 + 4 = 5$
$1 + 2 = 3$	$2 + 1 = 3$	$3 + 2 = 5$

4. $1 + \underline{4} = 5$ $4 + \underline{1} = 5$ $\underline{3} + 2 = 5$ $\underline{2} + 3 = 5$
   $2 + \underline{3} = 5$ $3 + \underline{2} = 5$ $\underline{5} + 0 = 5$ $\underline{4} + 1 = 5$
   $0 + \underline{5} = 5$ $5 + \underline{0} = 5$ $\underline{1} + 4 = 5$ $\underline{0} + 5 = 5$

5. a. 4, 5, 6   b. 6, 7, 8   c. 3, 4, 5
   d. 7, 8, 9   e. 5, 6, 7   f. 8, 9, 10

6. a. $2 + 3 = 5$ b. $1 + 2 = 3$ c. $3 + 1 = 4$
   d. $4 + 1 = 5$ e. $3 + 3 = 6$ f. $2 + 4 = 6$

Sums with 6, p. 36

1.

$0 + 6 = 6$	$6 + 0 = 6$
$1 + 5 = 6$	$5 + 1 = 6$
$2 + 4 = 6$	$4 + 2 = 6$
$3 + 3 = 6$	

3.
$1 + \underline{5} = 6$ $4 + \underline{2} = 6$ $\underline{4} + 2 = 6$ $\underline{3} + 3 = 6$
$2 + \underline{4} = 6$ $3 + \underline{3} = 6$ $\underline{6} + 0 = 6$ $\underline{5} + 1 = 6$
$6 + \underline{0} = 6$ $5 + \underline{1} = 6$ $\underline{2} + 4 = 6$ $\underline{1} + 5 = 6$

4. a. 6   b. 5   c. 6

5. a. $2 + 4 = 6$ b. $2 + 3 = 5$ c. $4 + 2 = 6$
   d. $3 + 3 = 6$ e. $1 + 5 = 6$ f. $5 + 1 = 6$
   g. $1 + 4 = 5$ h. $0 + 6 = 6$ i. $3 + 2 = 5$

6. a. 3, 4, 0, 2, 5, 1   b. 5, 2, 1, 6, 4, 3

7. 5, 5, 6, 6, 4, 6, 4, 6

11

Adding on a Number Line, p. 38

1. a. 7

b. 5

c. 9

d. 10

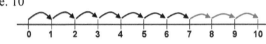

e. 10

f. 7

2. a. 3 + 3 = 6   b. 5 + 4 = 9   c. 2 + 8 = 10
   d. 6 + 1 = 7   e. 2 + 5 = 7   f. 4 + 4 = 8
   g. 6 + 2 = 8

3. a. 9

b. 5

c. 9

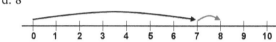

d. 8

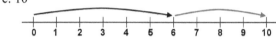

e. 10

f. 6

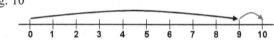

g. 10

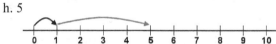

h. 5

4. a. 8 + 5 = 13   b. 9 + 3 = 12   c. 6 + 8 = 14

5. a. 8, 9   b. 6, 7   c. 7, 8   d. 9, 10
   e. 11,12   f. 13, 14   g. 14, 15   h. 12, 13

Sums with 7, p. 42

1.

0 + 7 = 7	7 + 0 = 7
1 + 6 = 7	6 + 1 = 7
2 + 5 = 7	5 + 2 = 7
3 + 4 = 7	4 + 3 = 7

2.

5 + <u>2</u> = 7   2 + <u>5</u> = 7   6 + <u>1</u> = 7   <u>4</u> + 3 = 7   <u>0</u> + 7 = 7
3 + <u>4</u> = 7   1 + <u>6</u> = 7   0 + <u>7</u> = 7   <u>5</u> + 2 = 7   <u>6</u> + 1 = 7
7 + <u>0</u> = 7   4 + <u>3</u> = 7   4 + <u>3</u> = 7   <u>1</u> + 6 = 7   <u>2</u> + 5 = 7

3. a. 6, 7   b. 7, 6   c. 7, 7   d. 7, 6

5.

a.	b.	c.
0 + 7 = 7	0 + 6 = 6	0 + 5 = 5
1 + 6 = 7	1 + 5 = 6	1 + 4 = 5
2 + 5 = 7	2 + 4 = 6	2 + 3 = 5
3 + 4 = 7	3 + 3 = 6	3 + 2 = 5

12

Sums with 7, cont.

6. a. 7 b. 6 c. 5 d. 5 e. 4
   f. 7 g. 3 h. 4 i. 6 j. 4
   k. 6 l. 2 m. 7 n. 7 o. 6

7. The use of pictures is optional. It helps many children, though, and in the future—even in algebra word problems—it is a good tactic for solving the problems. Some of the problems are simple addition problems; some are missing addend problems. To distinguish between addition and missing addend problems, you can ask: is the problem *asking* for the total, or do you *already know* the total? Is the problem asking how many there are together, or is it asking how many are missing?

   a. The problem asks for a total. The addition sentence is 3 + 6 = 9 goldfish.
   b. We know the total is 7. The picture would show initially two shirts, and then the child would draw some more so the total would be seven shirts. The red and other colors together make 7. The addition sentence for this problem is simply 2 + 5 = 7. Five of the T-shirts are not red.
   c. The problem asks for a total. 4 + 4 = 8 fish.

7. d. We know the total is 9. The picture would show six toy cars in the living room (perhaps inside a box). James has nine. The addition sentence is 6 + 3 = 9. Three cars are missing.
   e. We know the total is 7. The picture would have seven dolls and three hats. 3 + 4 = 7. She needs to find 4 more hats.
   f. The problem asks for a total. 2 + 4 = 6. She ate 6 cookies.

Puzzle Corner. There are many possible solutions; the ones below are just one possibility.

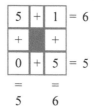

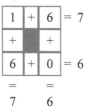

Sums with 8, p. 45

1.

🐚🐚🐚🐚🐚🐚🐚🐚 0 + 8 = 8	🐚🐚🐚🐚🐚🐚🐚🐚 8 + 0 = 8
🐚 | 🐚🐚🐚🐚🐚🐚🐚 1 + 7 = 8	🐚🐚🐚🐚🐚🐚🐚 | 🐚 7 + 1 = 8
🐚🐚 | 🐚🐚🐚🐚🐚🐚 2 + 6 = 8	🐚🐚🐚🐚🐚🐚 | 🐚🐚 6 + 2 = 8
🐚🐚🐚 | 🐚🐚🐚🐚🐚 3 + 5 = 8	🐚🐚🐚🐚🐚 | 🐚🐚🐚 5 + 3 = 8
🐚🐚🐚🐚 | 🐚🐚🐚🐚 4 + 4 = 8	

2.

$\underline{3} + 5 = 8$ $\underline{4} + 4 = 8$ $2 + \underline{6} = 8$
$\underline{8} + 0 = 8$ $\underline{2} + 6 = 8$ $5 + \underline{3} = 8$
$\underline{6} + 2 = 8$ $\underline{5} + 3 = 8$ $1 + \underline{7} = 8$

$3 + \underline{5} = 8$ $8 + \underline{0} = 8$
$7 + \underline{1} = 8$ $6 + \underline{2} = 8$
$4 + \underline{4} = 8$ $\underline{7} + 1 = 8$

4. a.

■ ■■■■■■■ 1 + 7 = 8
■■ ■■■■■■ 2 + 6 = 8
■■■ ■■■■■ 3 + 5 = 8
■■■■ ■■■■ 4 + 4 = 8

b.

■ ■■■■■■ 1 + 6 = 7
■■ ■■■■■ 2 + 5 = 7
■■■ ■■■■ 3 + 4 = 7
■■■■ ■■■ 4 + 3 = 7

c.

■ ■■■■■ 1 + 5 = 6
■■ ■■■■ 2 + 4 = 6
■■■ ■■■ 3 + 3 = 6
■■■■ ■■ 4 + 2 = 6

5.

a. 2 + 4 = 6	b. 2 + 3 = 5
c. 4 + 2 = 6	d. 5 + 3 = 8
e. 3 + 4 = 7	f. 2 + 2 = 4
g. 3 + 5 = 8	h. 2 + 6 = 8

6.

a.	b.	c.	d.
3 + 4 = 7 4 + 4 = 8	6 + 2 = 8 5 + 2 = 7	6 + 1 = 7 1 + 7 = 8	2 + 5 = 7 2 + 6 = 8
e.	f.	g.	h.
5 + 2 = 7 5 + 3 = 8	4 + 4 = 8 4 + 3 = 7	3 + 4 = 7 3 + 5 = 8	2 + 6 = 8 2 + 5 = 7

7.

a. 4   b. 6   c. 3   d. 7   e. 5
   + 2   + 2   + 3   + 1   + 2
   ‾‾‾   ‾‾‾   ‾‾‾   ‾‾‾   ‾‾‾
    6   8   6   8   7

f. 1   g. 6   h. 4   i. 5   j. 3
   + 2   + 1   + 3   + 1   + 2
   ‾‾‾   ‾‾‾   ‾‾‾   ‾‾‾   ‾‾‾
    3   7   7   6   5

8. a. 7 = 7   b. 7 < 8   c. 6 > 4   d. 10 = 10
   e. 8 > 4   f. 2 = 2   g. 0 = 0   h. 8 > 7
   i. 4 = 4   j. 1 < 5   k. 6 < 8   l. 2 > 0

13

Adding Many Numbers, p. 48

1. a. 9, 9   b. 9, 9   c. 10, 10

2. a. 10   b. 8   c. 8   d. 9   e. 7
   f. 10   g. 10   h. 9   i. 10   j. 9

3. a. She has 7 flowers. 2 + 3 + 2 = 7
   b. She used 9 chairs. 3 + 3 + 3 = 9
   c. Four were missing. 10 − 6 = 4 or 6 + 4 = 10.

4. a. true   b. false   c. false   d. true

5.

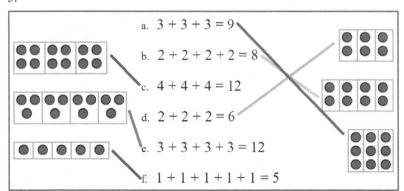

6. a. 1 + 2 + 3 = 6
   b. 1 + 1 + 4 = 6
   c. 2 + 1 + 2 + 3 = 8
   d. 3 + 3 + 3 + 1 = 10

7. a. 8, 8, 10   b. 9, 7, 10   c. 10, 5, 10

Addition Practice 2, p. 51

1.

a. 4 + 4 = 8	b. 4 + 3 = 7	c. 2 + 4 = 6
6 + 2 = 8	5 + 2 = 7	1 + 6 = 7

2. a. 4   b. 6   c. 8   d. 10   e. 0   f. 2

3.

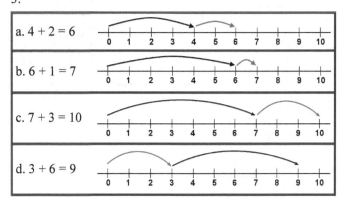

4. a. 9, 9   b. 7, 7   c. 8, 8   d. 5, 5

5.
a. Add 1

5 + 1 = 6
6 + 1 = 7
7 + 1 = 8
8 + 1 = 9
9 + 1 = 10

b. Add 2

2 + 2 = 4
3 + 2 = 5
4 + 2 = 6
5 + 2 = 7
6 + 2 = 8

c. Add 3

2 + 3 = 5
3 + 3 = 6
4 + 3 = 7
5 + 3 = 8
6 + 3 = 9

6.

+	1	2	3
1	2	3	4
2	3	4	5
3	4	5	6

+	1	2	3
4	5	6	7
5	6	7	8
6	7	8	9

1.

$0 + 9 = 9$	$9 + 0 = 9$
$1 + 8 = 9$	$8 + 1 = 9$
$2 + 7 = 9$	$7 + 2 = 9$
$3 + 6 = 9$	$6 + 3 = 9$
$4 + 5 = 9$	$5 + 4 = 9$

2. $\underline{1} + 8 = 9$ $\underline{5} + 4 = 9$ $2 + \underline{7} = 9$ $3 + \underline{6} = 9$ $7 + \underline{2} = 9$
   $\underline{7} + 2 = 9$ $\underline{3} + 6 = 9$ $9 + \underline{0} = 9$ $6 + \underline{3} = 9$ $\underline{8} + 1 = 9$
   $\underline{2} + 7 = 9$ $\underline{6} + 3 = 9$ $0 + \underline{9} = 9$ $4 + \underline{5} = 9$ $\underline{4} + 5 = 9$

4. a. 9 b. 3 c. 1 d. 5 e. 7 f. 6

5.   a.

$1 + 6 = 7$
$2 + 5 = 7$
$3 + 4 = 7$
$4 + 3 = 7$

b.

$1 + 7 = 8$
$2 + 6 = 8$
$3 + 5 = 8$
$4 + 4 = 8$

c.

$1 + 8 = 9$
$2 + 7 = 9$
$3 + 6 = 9$
$4 + 5 = 9$

6. a. $\frac{2 + 5}{7}$ b. $\frac{1 + 6}{7}$ c. $\frac{4 + 4}{8}$ d. $\frac{7 + 1}{8}$ e. $\frac{7 + 2}{9}$ f. $\frac{3 + 5}{8}$ g. $\frac{4 + 2}{6}$ h. $\frac{3 + 4}{7}$ i. $\frac{1 + 5}{6}$ j. $\frac{4 + 5}{9}$

7. a. We know the total is 5. 2 + 3 = 5. She needs 3 more eggs.
   b. The total is 8. 4 + 4 = 8. So, 4 crayons are missing.
   c. The problem asks for the total. 5 + 5 = 10. Jenny and Penny have 10 goldfish.
      With Betty's fish, 10 + 3 = 13 goldfish altogether.
   d. You cannot buy the doll with two dollars. 8 + 2 = 10. You have $10 together. You can buy the doll together.
   e. The problem asks for the total. 2 + 6 = 8. There are 8 red chairs in the house.
   f. Two dollars more. 5 + 2 = 7.
   g. Two dollars more. 8 + 2 = 10.
   h. Nine dollars. 5 + 4 = 9.

8.

a. 5 + 2 4	b. 4 + 4 7	c. 2 1 + 1	d. 7 3 + 6
↓ ↓	↓ ↓	↓ ↓	↓ ↓
7 > 4	8 > 7	2 = 2	7 < 9

9. a. $1 + 4 > 3$ b. $2 + 2 < 5$ c. $2 > 0 + 0$ d. $7 < 5 + 3$
   e. $4 + 4 < 9$ f. $3 + 5 > 6$ g. $7 < 6 + 2$ h. $8 > 3 + 4$

Sums with 9, cont.

Puzzle Corner. There are many solutions. These are just examples.

1	+	8	=	9
+	■	+		
9	+	0	=	9
=		=		
10		8		

1	+	8	=	9
+	■	+		
8	+	0	=	8
=		=		
9		8		

Sums with 10, p. 57

1.

〰〰〰〰〰〰〰〰〰〰 0 + 10 = 10	〰〰〰〰〰〰〰〰〰〰 10 + 0 = 10
〰 | 〰〰〰〰〰〰〰〰〰 1 + 9 = 10	〰〰〰〰〰〰〰〰〰 | 〰 9 + 1 = 10
〰〰 | 〰〰〰〰〰〰〰〰 2 + 8 = 10	〰〰〰〰〰〰〰〰 | 〰〰 8 + 2 = 10
〰〰〰 | 〰〰〰〰〰〰〰 3 + 7 = 10	〰〰〰〰〰〰〰 | 〰〰〰 7 + 3 = 10
〰〰〰〰 | 〰〰〰〰〰〰 4 + 6 = 10	〰〰〰〰〰〰 | 〰〰〰〰 6 + 4 = 10
〰〰〰〰〰 | 〰〰〰〰〰 5 + 5 = 10	

3.

$\underline{4}$ + 6 = 10	$\underline{6}$ + 4 = 10	1 + $\underline{9}$ = 10
$\underline{7}$ + 3 = 10	$\underline{5}$ + 5 = 10	7 + $\underline{3}$ = 10
$\underline{2}$ + 8 = 10	$\underline{1}$ + 9 = 10	2 + $\underline{8}$ = 10

6 + $\underline{4}$ = 10	3 + $\underline{7}$ = 10
9 + $\underline{1}$ = 10	4 + $\underline{6}$ = 10
5 + $\underline{5}$ = 10	8 + $\underline{2}$ = 10

4. a.

| ■■ ■■■■■■■■
 2 + 8 = 10 |
| ■■■ ■■■■■■■
 3 + 7 = 10 |
| ■■■■ ■■■■■■
 4 + 6 = 10 |
| ■■■■■ ■■■■■
 5 + 5 = 10 |

b.

| ■■ ■■■■■■■
 2 + 7 = 9 |
| ■■■ ■■■■■■
 3 + 6 = 9 |
| ■■■■ ■■■■■
 4 + 5 = 9 |
| ■■■■■ ■■■■
 5 + 4 = 9 |

c.

| ■■ ■■■■■■
 2 + 6 = 8 |
| ■■■ ■■■■■
 3 + 5 = 8 |
| ■■■■ ■■■■
 4 + 4 = 8 |
| ■■■■■ ■■■
 5 + 3 = 8 |

5.
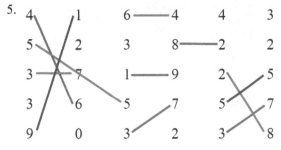

6. a. 6 < 7   b. 10 > 8   c. 6 < 8   d. 10 = 10
   e. 8 > 6   f. 5 = 5   g. 9 > 8   h. 5 < 10

7.

a. 1 + 9 > 9	b. 4 + 4 < 9	c. 6 < 5 + 2	d. 9 = 5 + 4
e. 5 + 5 = 10	f. 3 + 5 > 7	g. 10 > 6 + 3	h. 7 < 7 + 1

8.

a.	b.	c.
0 + 10 = 10	6 + 4 = 10	7 + 3 = 10
5 + 5 = 10	2 + 8 = 10	2 + 8 = 10
9 + 1 = 10	4 + 6 = 10	1 + 9 = 10

9.

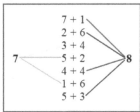
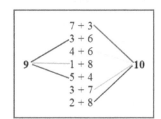

10. a. The problem asks for the total.
   There are 3 + 7 = 10 birds now.
   b. We already know the total is 7. 3 + 4 = 7
   There are four books that she hasn't read.
   c. We already know the total is 10.   4 + 6 = 10
   Six of the dolls are not in her room.
   d. The problem asks for the total. 3 + 3 = 6
   They have 6 cars.
   e. We already know the total is 10.   6 + 4 = 10
   Four are missing.
   f. The problem asks for the total. 2 + 5 = 7 birds.
   g. We already know the total is 10. 5 + 5 = 10
   Jessica has 5 books.
   h. We already know the total is 10. 2 + 8 = 10.
   There are 8 dolls on the top shelf.

1.

a. $4 + 1 = 5$	b. $7 < 4 + 4$	c. $6 > 2 + 3$
d. $2 + 5 = 7$	e. $5 = 5 + 0$	f. $10 = 5 + 5$
g. $2 + 2 > 3$	h. $9 = 9$	i. $2 < 2 + 2$

2.

a. $5 < 6$	b. $4 < 5$	c. $7 > 6$	d. $4 > 3$
e. $9 > 7$	f. $3 < 5$	g. $7 > 6$	h. $2 < 3$

3.

a. $2 + 4 = 6$ b. $1 + 4 < 6$ c. $4 + \underline{1 \text{ or } 2} < 7$

d. $2 + \underline{5 \text{ or } 6} > 6$ e. $1 + 5 = 6$ f. $1 + 9 > 9$

g. $10 = 2 + 8$ h. $3 + 2 < 7$ i. $4 + \underline{5 \text{ or } 6} > 8$

4.

a. $4 + 3 > 5$	b. $7 + 1 < 9$	c. $4 < 4 + 2$
d. $2 + 5 < 8$	e. $3 + 4 > 6$	f. $6 = 3 + 3$
g. $8 + 2 = 10$	h. $9 + 2 > 9$	i. $2 < 2 + 1$

5. a. $7 + 3 = 2 + 8$ b. $1 + 1 < 1 + 4$ c. $4 < 1 + 4$

    d. $5 + 4 = 4 + 5$ e. $2 + 5 > 2 + 2$ f. $3 < 3 + 1$

    g. $2 + 4 > 2 + 1$ h. $10 + 0 = 0 + 10$ i. $0 = 0 + 0$

6.

a. true	d. false
b. false	e. true
c. false	f. false

7. Answers will vary. For example:

    a. $10 = 9 + 1 = 10$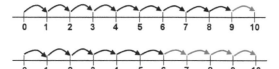

    b. $10 = 6 + 4 = 10$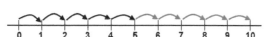

    c. $10 = 5 + 5 = 10$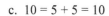

8.

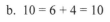

+	1	2	3	4	5	6	7
0	1	2	3	4	5	6	7
1	2	3	4	5	6	7	8
2	3	4	5	6	7	8	9
3	4	5	6	7	8	9	10
4	5	6	7	8	9	10	11
5	6	7	8	9	10	11	12

1. Answers will vary.

5 = 1 + 4	6 = 0 + 6
5 = 2 + 3	6 = 1 + 5
5 = 0 + 5	6 = 2 + 4
5 = 3 + 2	6 = 3 + 3

9. Answers will vary.

9 = 9 + 0	10 = 0 + 10
9 = 1 + 8	10 = 1 + 9
9 = 7 + 2	10 = 2 + 8
9 = 3 + 6	10 = 3 + 7
9 = 4 + 5	10 = 4 + 6
9 = 2 + 7	10 = 5 + 5

2.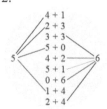

3.

4 + 2 = 6	6 + 0 = 6
2 + 3 = 5	0 + 5 = 5
1 + 4 = 5	3 + 3 = 6
6 + 0 = 6	4 + 2 = 6
5 + 1 = 6	1 + 4 = 5

4.

2 + 2 < 5	4 + 4 > 5	6 = 2 + 4
2 + 3 = 5	5 + 5 > 5	6 < 2 + 5
2 + 4 > 5	5 + 0 = 5	6 < 2 + 6

5. Answers will vary.

7 = 0 + 7	8 = 0 + 8
7 = 1 + 6	8 = 1 + 7
7 = 2 + 5	8 = 2 + 6
7 = 3 + 4	8 = 3 + 5
7 = 4 + 3	8 = 4 + 4
7 = 5 + 2	8 = 5 + 3

10.

11.

8 + 2 = 10	3 + 6 = 9
5 + 4 = 9	7 + 3 = 10
2 + 7 = 9	3 + 6 = 9
5 + 5 = 10	7 + 2 = 9
6 + 4 = 10	4 + 6 = 10

12.

2 + 6 < 9	6 + 6 > 10	10 < 10 + 4
4 + 6 > 9	5 + 5 = 10	10 = 10 + 0
3 + 6 = 9	4 + 4 < 10	10 < 10 + 7

13.

a.	b.	c.
8 + 1 = 9	4 + 1 + 1 = 6	5 + 2 + 0 + 0 = 7
6 + 2 = 8	8 + 2 + 0 = 10	4 + 3 + 1 + 2 = 10
1 + 7 = 8	1 + 3 + 6 = 10	1 + 2 + 2 + 1 = 6
3 + 4 = 7	2 + 2 + 4 = 8	2 + 3 + 1 + 3 = 9

6.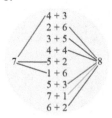

7.

5 + 2 = 7	4 + 4 = 8
3 + 4 = 7	3 + 4 = 7
2 + 6 = 8	3 + 5 = 8
5 + 3 = 8	7 + 1 = 8
6 + 1 = 7	5 + 2 = 7

14.

+	2	4	3	6	7	5	8
1	3	5	4	7	8	6	9
3	5	7	6	9	10	8	11
4	6	8	7	10	11	9	12
2	4	6	5	8	9	7	10

8.

3 + 3 < 7	6 + 1 = 7	8 < 6 + 4
4 + 3 = 7	6 + 6 > 7	8 = 4 + 4
5 + 3 > 7	6 + 4 > 7	8 < 5 + 4

Puzzle corner:

a. △ = 3 and ☐ = 5

b. ☐ = 5 and △ = 2

c. ☐ = 3 and △ = 2

Chapter 2: Subtraction Within 0-10

Subtraction is "Taking Away", p. 71

1.

a. $10 - 4 = 6$	b. $6 - 6 = 0$	c. $8 - 2 = 6$
d. $8 - 7 = 1$	e. $8 - 6 = 2$	f. $7 - 3 = 4$
g. $9 - 3 = 6$	h. $6 - 2 = 4$	i. $10 - 7 = 3$
j. $9 - 5 = 4$	k. $8 - 4 = 4$	l. $10 - 5 = 5$
m. $6 - 3 = 3$	n. $6 - 5 = 1$	o. $5 - 5 = 0$

2. a. 4 b. 3 c. 2 d. 3 e. 3 f. 4 g. 6 h. 6 i. 5 j. 6 k. 4 l. 4

3. a. 5 b. 4 c. 3 d. 5 e. 9 f. 2 g. 3 h. 4 i. 6 j. 4 k. 4 l. 2

4. a. $8 - 4 = 4$ b. $7 - 4 = 3$
   c. $7 - 1 = 6$ d. $9 - 6 = 3$
   e. $9 - 7 = 2$ f. $10 - 2 = 8$

Count Down to Subtract, p. 74

1.

a. $7 - 2 = \underline{5}$

b. $8 - 4 = \underline{4}$

c. $6 - 5 = \underline{1}$

d. $9 - 3 = \underline{6}$

e. $10 - 3 = \underline{7}$

2. a. $6 - 4 = 2$ b. $4 - 1 = 3$ c. $7 - 5 = 2$ d. $9 - 4 = 5$

3.

a. $10 - 5 = \underline{5}$

b. $7 - 6 = \underline{1}$

c. $4 - 4 = \underline{0}$

d. $8 - 4 = \underline{4}$

e. $10 - 1 = \underline{9}$

f. $7 - 5 = \underline{2}$

4. a. <u>4</u>, 5, <u>6</u>   b. <u>1</u>, 2, <u>3</u>   c. <u>7</u>, 8, <u>9</u>
   d. <u>5</u>, 6 , <u>7</u>   e. <u>3</u>, 4 , <u>5</u>   f. <u>8</u>, 9 , <u>10</u>

5. a. <u>5</u>, <u>6</u>, 7   b. <u>2</u>, <u>3</u>, 4   c. <u>8</u>, <u>9</u>, 10
   d. <u>4</u>, <u>5</u>, 6   e. <u>0</u>, <u>1</u>, 2   f. <u>6</u>, <u>7</u>, 8

6. a. 5, 7   b. 4, 6   c. 8, 2   d. 3, 9

7. a. 4, 3   b. 7, 6   c. 5, 4   d. 8, 7

8. All of these situations are of the type "taking away," "going away," and so on.
   a. $7 - 3 = 4$;   There are 4 birds left in the tree.
   b. $10 - 4 = 6$;   There are still six silver plates in the cupboard.
   c. $9 - 4 = 5$;   Five girls kept jumping rope.
   d. $10 - 5 = 5$;   Josh left five of his toy cars at home.
   e. $8 - 3 = 5$;   Fanny still has five puzzles to play with.
   f. $6 - 6 = 0$;   Tina does not have any flowers left.

9. a. 12, 10   b. 15, 13   c. 8, 7   d. 15, 14   e. 11, 10   f. 11, 10

Subtraction and Addition in the Same Picture, p. 78

1.

a. $3 + 4 = 7$ $7 - 4 = 3$ or $7 - 3 = 4$	b. $4 + 2 = 6$ $6 - 2 = 4$ or $6 - 4 = 2$
c. $1 + 4 = 5$ $5 - 4 = 1$ or $5 - 1 = 4$	d. $5 + 1 = 6$ $6 - 1 = 5$ or $6 - 5 = 1$
e. $5 + 3 = 8$ $8 - 3 = 5$ or $8 - 5 = 3$	f. $3 + 3 = 6$ $6 - 3 = 3$

2.

a.	$5 + 4 = 9$ $9 - 4 = 5$ or $9 - 5 = 4$	b.	$3 + 6 = 9$ $9 - 6 = 3$ or $9 - 3 = 6$
c.	$5 + 5 = 10$ $10 - 5 = 5$	d.	$6 + 6 = 12$ $12 - 6 = 6$

3.

○○○○○● a. $7 + 1 = 8$ $8 - 1 = 7$	○○○○○○●●● b. $6 + 3 = 9$ $9 - 3 = 6$
○○●●● c. $2 + 3 = 5$ $5 - 3 = 2$	○○●●●●● d. $2 + 5 = 7$ $7 - 2 = 5$
○○○○○○○●●●● e. $7 + 4 = 11$ $11 - 4 = 7$	○○○●●● f. $3 + 3 = 6$ $6 - 3 = 3$

4. a. $10 - 4 = 6$ b. $8 - 5 = 3$ c. $6 - 2 = 4$ d. $5 - 4 = 1$
   e. $5 - 1 = 4$ f. $6 - 3 = 3$ g. $11 - 7 = 4$ h. $11 - 6 = 5$

5. Either subtraction sentence could be correct, depending on which circles the student colored.

○○○○○○○○● a. $9 + 1 = 10$ $10 - 9 = 1$ or $10 - 1 = 9$	○○○○○○○●● b. $7 + 2 = 9$ $9 - 7 = 2$ or $9 - 2 = 7$
○○○○○○○○○○ ●●●● c. $10 + 4 = 14$ $14 - 10 = 4$ or $14 - 4 = 10$	○○○○○○○○○○ ●● d. $10 + 2 = 12$ $12 - 10 = 2$ or $12 - 2 = 10$

6.

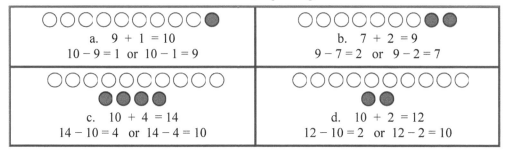

○○○○○⊘⊘⊘ a. $9 - 4 = 5$ $5 + 4 = 9$	○○○○○⊘⊘⊘⊘⊘ b. $10 - 5 = 5$ $5 + 5 = 10$
○○○⊘⊘⊘⊘⊘ c. $8 - 5 = 3$ $3 + 5 = 8$	○○○○⊘⊘⊘⊘ d. $8 - 4 = 4$ $4 + 4 = 8$
○○○⊘⊘⊘⊘ e. $7 - 4 = 3$ $4 + 3 = 7$	○⊘⊘⊘⊘⊘⊘⊘⊘ f. $9 - 8 = 1$ $1 + 8 = 9$

Puzzle Corner.

3	$>$	$3 - 1$	$6 + 5$	$>$	6
$9 - 7$	$>$	$8 - 7$	$6 - 4$	$<$	$2 + 3$
$5 + 2$	$<$	$8 + 2$	$10 - 1$	$>$	$10 - 3$
$10 - 2$	$>$	$8 - 2$	$10 + 0$	$=$	$10 - 0$

10	$>$	$10 - 1$
$8 - 5$	$<$	$5 + 3$
$7 - 4$	$=$	$8 - 5$
$8 - 1$	$<$	$8 + 1$

1.

4 − 0 = 4	6 − 0 = 6	5 − 0 = 5
4 − 1 = 3	6 − 1 = 5	5 − 1 = 4
4 − 2 = 2	6 − 2 = 4	5 − 2 = 3
4 − 3 = 1	6 − 3 = 3	5 − 3 = 2
4 − 4 = 0	6 − 4 = 2	5 − 4 = 1
	6 − 5 = 1	5 − 5 = 0
	6 − 6 = 0	

You cannot do a subtraction problem in whole numbers (0, 1, 2, 3, etc.) when the second number is <u>bigger / larger / greater</u> than the first number.

2.

a.	b.	c.	d.
7 − 1 = 6	9 − 1 = 8	10 − 1 = 9	12 − 1 = 11
7 − 2 = 5	9 − 2 = 7	10 − 2 = 8	12 − 2 = 10
7 − 3 = 4	9 − 3 = 6	10 − 3 = 7	12 − 3 = 9

3.

a.	b.	c.
7 − 0 = 7	10 − 5 = 5	8 − 2 = 6
7 − 1 = 6	9 − 5 = 4	7 − 2 = 5
7 − 2 = 5	8 − 5 = 3	6 − 2 = 4
7 − 3 = 4	7 − 5 = 2	5 − 2 = 3
7 − 4 = 3	6 − 5 = 1	4 − 2 = 2
7 − 5 = 2	5 − 5 = 0	3 − 2 = 1
7 − 6 = 1		2 − 2 = 0
7 − 7 = 0		

4.

4 − 0   7 − 7   <s>5 − 6</s>   <s>3 − 6</s>   4 − 4   <s>3 − 10</s>

     10 − 1   <s>3 − 4</s>   <s>2 − 4</s>   4 − 3

5. a. false   b. true   c. false   d. false   e. false   f. true.

6. (Note to the teacher: The pictures that the child is to draw need only be rough sketches. For example, he or she could draw tiny rectangles or circles for money, a stick man for the person, and a line, a box, or a circle (with a price tag) for the object to buy. If children can draw a picture, it usually means that they have understood the problem. Drawing a picture will be a very important strategy for many kinds of math problems at higher levels, too, so it is good that students learn to use it as early as possible.)

a. Jennie cannot buy the doll. She needs two dollars more.   b. Jessie can buy the ball. He will have three dollars left.
c. Lola can buy the Lego set and have three dollars left.   d. Marvin cannot buy the book. He needs one dollar more.
e. Jack cannot buy the set. He needs two dollars more.   f. Mary can buy the car. She will have five dollars left.
g. Faye can buy the game and have eight dollars left.   h. He cannot buy the game. He needs five dollars more.

When Can You Subtract?, cont.

7. a.
$$\begin{array}{r} 10 \\ -\ 3 \\ \hline 7 \end{array}$$
b.
$$\begin{array}{r} 8 \\ -\ 7 \\ \hline 1 \end{array}$$
c.
$$\begin{array}{r} 6 \\ -\ 5 \\ \hline 1 \end{array}$$
d.
$$\begin{array}{r} 8 \\ -\ 6 \\ \hline 2 \end{array}$$
e.
$$\begin{array}{r} 8 \\ -\ 0 \\ \hline 8 \end{array}$$

f.
$$\begin{array}{r} 7 \\ -\ 7 \\ \hline 0 \end{array}$$
g.
$$\begin{array}{r} 7 \\ -\ 6 \\ \hline 1 \end{array}$$
h.
$$\begin{array}{r} 6 \\ -\ 6 \\ \hline 0 \end{array}$$
i.
$$\begin{array}{r} 6 \\ -\ 1 \\ \hline 5 \end{array}$$
j.
$$\begin{array}{r} 9 \\ -\ 4 \\ \hline 5 \end{array}$$

k.
$$\begin{array}{r} 10 \\ -\ 8 \\ \hline 2 \end{array}$$
l.
$$\begin{array}{r} 4 \\ -\ 0 \\ \hline 4 \end{array}$$
m.
$$\begin{array}{r} 6 \\ -\ 4 \\ \hline 2 \end{array}$$
n.
$$\begin{array}{r} 7 \\ -\ 2 \\ \hline 5 \end{array}$$
o.
$$\begin{array}{r} 9 \\ -\ 3 \\ \hline 6 \end{array}$$

8.

10 − 6	6 − 7	5 − 6	5 − 10	7 − 3
1 − 6	10 − 5	9 − 10	8 − 3	1 − 3
7 − 9	1 − 2	10 − 0	3 − 5	6 − 9
4 − 8	6 − 1	0 − 9	9 − 4	7 − 8
8 − 4	0 − 4	2 − 6	6 − 9	9 − 5

Two Subtractions from One Addition, p. 86

1.

a. $1 + 3 = 4$ $4 - 3 = 1$ or $4 - 1 = 3$	b. $2 + 3 = 5$ $5 - 3 = 2$ or $5 - 2 = 3$
c. $4 + 5 = 9$ $9 - 4 = 5$ or $9 - 5 = 4$	d. $5 + 2 = 7$ $7 - 5 = 2$ or $7 - 2 = 5$
e. $2 + 4 = 6$ $6 - 2 = 4$ or $6 - 4 = 2$	f. $1 + 6 = 7$ $7 - 1 = 6$ or $7 - 6 = 1$
g. $4 + 5 = 9$ $9 - 4 = 5$ or $9 - 5 = 4$	h. $3 + 7 = 10$ $10 - 3 = 7$ or $10 - 7 = 3$
i. $3 + 3 = 6$ $6 - 3 = 3$ or $6 - 3 = 3$	j. $5 + 3 = 8$ $8 - 5 = 3$ or $8 - 3 = 5$
k. $2 + 6 = 8$ $8 - 2 = 6$ or $8 - 6 = 2$	l. $8 + 4 = 12$ $12 - 8 = 4$ or $12 - 4 = 8$

2.

XXXXXX X a. $6 + 1 = 7$ $7 - 1 = 6$ $7 - 6 = 1$	XXXX XXX b. $4 + 3 = 7$ $7 - 4 = 3$ $7 - 3 = 4$
XXXXXX X c. $7 + 1 = 8$ $8 - 1 = 7$ $8 - 7 = 1$	XXXX XXXXX d. $5 + 4 = 9$ $9 - 5 = 4$ $9 - 4 = 5$
XX XXXXX e. $2 + 6 = 8$ $8 - 2 = 6$ $8 - 6 = 2$	XXX XXXXX f. $3 + 5 = 8$ $8 - 3 = 5$ $8 - 5 = 3$
X XXXXXXX g. $1 + 8 = 9$ $9 - 1 = 8$ $9 - 8 = 1$	XXXX XXXXX h. $5 + 5 = 10$ $10 - 5 = 5$ $10 - 5 = 5$

Two Parts—One Total, p. 89

Teaching box:	There are five blue marbles and some white marbles in a bag. <u>Four are white.</u>	 $5 + 4 = 9$ $9 - 5 = 4$

1. a. There are five blue flowers.

$5 + 5 = 10$
$10 - 5 = 5$

  b. There are five girls.

$4 + 5 = 9$
$9 - 4 = 5$

  c. Two of the socks are black.

$8 + 2 = 10$
$10 - 8 = 2$

  d. Six chairs were still standing
    upright on the lawn.

$2 + 6 = 8$
$8 - 2 = 6$

2. Answers will vary. For example:

  a. $8 - 2 = 6$. Gladys drew 8 circles and colored two of them yellow. The rest she colored red. There are six red circles.
  b. $7 - 3 = 4$. Together Joe and his sister had seven toys. Joe had three cars. Joe's sister had four dolls.

3. a. $2 + 1 + 3 = 6$ b. $3 + 2 + 2 = 7$ c. $1 + 4 + 3 = 8$ d. $3 + 2 + 4 = 9$

4.

a. $3 + 2 + 3 = 8$	b. $1 + 5 + 4 = 10$

5. a. $2 + 2 + 6 = 10$ There were six yellow roses
  b. $1 + 2 + 4 = 7$ Four birds were brown.
  c. $2 + 2 + 5 = 9$ Five of her pencils are short.

Fact Families, p. 92

1.

a. $1 + 5 = 6$ $5 + 1 = 6$ $6 - 1 = 5$ $6 - 5 = 1$	b. $3 + 5 = 8$ $5 + 3 = 8$ $8 - 3 = 5$ $8 - 5 = 3$
c. $3 + 6 = 9$ $6 + 3 = 9$ $9 - 6 = 3$ $9 - 3 = 6$	d. $7 + 3 = 10$ $3 + 7 = 10$ $10 - 3 = 7$ $10 - 7 = 3$

Fact Families, cont.

2. Answers will vary. Some examples:

a. 7	b. 7
● / ⣿	⣤ / ⣤
$1 + 6 = 7$	$3 + 4 = 7$
$6 + 1 = 7$	$4 + 3 = 7$
$7 - 1 = 6$	$7 - 3 = 3$
$7 - 6 = 1$	$7 - 4 = 4$
c. 7	d. 7
⣿ / ⣀	⣿⣀ /
$5 + 2 = 7$	$7 + 0 = 7$
$2 + 5 = 7$	$0 + 7 = 7$
$7 - 2 = 5$	$7 - 7 = 7$
$7 - 5 = 2$	$7 - 0 = 0$

3.

a. Correct	d. Should be $\underline{5} - 4 = 1$
b. Should be $\underline{5} - 4 = 1$	e. Should be $10 - 8 = \underline{2}$
c. Should be $5 - \underline{3} = 2$	f. Correct

4.

a. Numbers: 5, 3, 2	b. Numbers: 9, 4, 5
$2 + 3 = 5$	$4 + 5 = 9$
$3 + 2 = 5$	$5 + 4 = 9$
$5 - 3 = 2$	$9 - 4 = 5$
$5 - 2 = 3$	$9 - 5 = 4$
c. Numbers: 4, 0, 4	d. Numbers: 10, 3, 7
$4 + 0 = 4$	$3 + 7 = 10$
$0 + 4 = 4$	$7 + 3 = 10$
$4 - 0 = 4$	$10 - 7 = 3$
$4 - 4 = 0$	$10 - 3 = 7$
e. Numbers: 10, 2, 8	f. Numbers: 6, 0, 6
$2 + 8 = 10$	$6 + 0 = 6$
$8 + 2 = 10$	$0 + 6 = 6$
$10 - 8 = 2$	$6 - 0 = 6$
$10 - 2 = 8$	$6 - 6 = 0$

5.

a. Numbers: 10, 5, 5	b. Numbers: 9, 1, 8
$5 + 5 = 10$	$1 + 8 = 9$
$5 + 5 = 10$	$8 + 1 = 9$
$10 - 5 = 5$	$9 - 8 = 1$
$10 - 5 = 5$	$9 - 1 = 8$
c. Numbers: 6, 3, 3	d. Numbers: 7, 1, 6
$3 + 3 = 6$	$1 + 6 = 7$
$3 + 3 = 6$	$6 + 1 = 7$
$6 - 3 = 3$	$7 - 6 = 1$
$6 - 3 = 3$	$7 - 1 = 6$

Fact Families, cont.

Puzzle Corner.

$9 - 4 = 5$	$5 - 1 = 4$	$1 + 4 = 5$
$5 - 2 = 3$	$7 - 6 = 1$	$1 + 2 = 3$
$8 - 7 = 1$	$10 - 5 = 5$	$3 + 7 = 10$

How Many More?, p. 96

1. a. 2 more, 2 fewer   b. 3 more, 3 fewer   c. 4 more, 4 fewer   d. 5 more, 5 fewer   e. 7 more, 7 fewer   f. 5 more, 5 fewer

2. a. Jane has 9 marbles.   b. Mary has 6 marbles.   c. Eric has 2 marbles.   d. Jane has 2 marbles   e. Bill has 6 marbles.
   f. Liz has 5 marbles.   g. Ed has 8 marbles.   h. Mary has 7 marbles.   i. Sue has 3 marbles.   j. Mary has 9 marbles.

3. Check the student's work.

4. a. 2 more.   b. 3 years.   c. 2 years.   d. 5 fewer dolls.

"How Many More" Problems and Differences, p. 99

1. a. $1 + 3 = 4$ b. $3 + 3 = 6$ c. $3 + 4 = 7$ d. $4 + 6 = 10$ e. $2 + 6 = 8$ f. $1 + 4 = 5$

2. a. $2 + 5 = 7$ b. $4 + 2 = 6$ c. $7 + 1 = 8$ d. $5 + 2 = 7$ e. $3 + 4 = 7$ f. $3 + 3 = 6$

3. a. 4 steps   b. 3 steps   c. 4 steps   d. 0 steps   e. 9 steps

4.

From	8	4	1	3	6	10	8	9
To	10	10	9	1	5	5	12	15
Difference	2	6	8	2	1	5	4	6

5.

a. from 3 to 5	b. from 1 to 5	c. from 2 to 7
2 steps	4 steps	5 steps
$3 + 2 = 5$	$1 + 4 = 5$	$2 + 5 = 7$

6.

a. from 6 to 9	b. from 4 to 8	c. from 8 to 9	d. from 2 to 6
3 steps	4 steps	1 step	4 steps
$6 + 3 = 9$	$4 + 4 = 8$	$8 + 1 = 9$	$2 + 4 = 6$

7. a. $7 + 3 = 10$. Jill has 3 more.   b. $4 + 3 = 7$. Al has 3 more.
   c. $4 + 2 = 6$. Ann has 2 more.   d. $2 + 7 = 9$. Hannah has 7 more.
   e. $10 + 1 = 11$. Britney has 1 more.   f. $5 + 5 = 10$. Don has 5 more.

8. a. $2 + 8 = 10$. There are 10 tapes. $2 + 6 = 8$ OR $8 - 2 = 6$. There are 6 more tapes on the shelf than on the table.
   b. $5 + 4 = 9$. There are now 9 birds now in the oak tree. $5 + 4 = 9$ OR $9 - 5 = 4$.
      There are 4 more birds in the oak tree.
   c. $9 + 1 = 10$ OR $10 - 9 = 1$. Joe has 1 more car than Jason. $2 + 7 = 9$ OR $9 - 2 = 7$.
      Jason has 7 more cars than Brenda.

1.

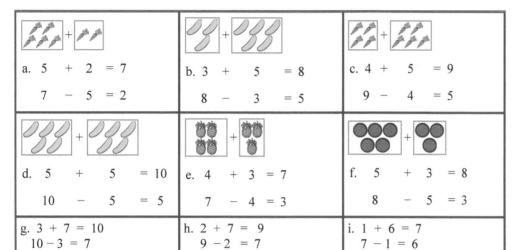

a. 5 + 2 = 7

7 − 5 = 2

b. 3 + 5 = 8

8 − 3 = 5

c. 4 + 5 = 9

9 − 4 = 5

d. 5 + 5 = 10

10 − 5 = 5

e. 4 + 3 = 7

7 − 4 = 3

f. 5 + 3 = 8

8 − 5 = 3

g. 3 + 7 = 10
   10 − 3 = 7

h. 2 + 7 = 9
   9 − 2 = 7

i. 1 + 6 = 7
   7 − 1 = 6

2.

a. 2 + 6 = 8	b. 1 + 8 = 9	c. 7 + 3 = 10	d. 6 + 3 = 9
8 − 2 = 6	9 − 1 = 8	10 − 3 = 7	9 − 3 = 6

3.

a. 1 + 6 = 7	b. 2 + 7 = 9	c. 1 + 9 = 10
7 − 1 = 6	9 − 2 = 7	10 − 1 = 9
d. 5 + 3 = 8	e. 8 + 2 = 10	f. 6 + 3 = 9
8 − 3 = 5	10 − 2 = 8	9 − 3 = 6

4. a. 2 + 6 = 8. They ate 8 carrots.   b. 3 + 4 = 7. The baby used 7 blocks.
   c. 5 + 2 = 7. There are now 7 lambs.   d. 4 + 4 = 8 or 8 − 4 = 4. She needs 4 dollars.

5. a. 8 − 6 = 2, 6 + 2 = 8   b. 10 − 9 = 1, 9 + 1 = 10   c. 9 − 7 = 2, 7 + 2 = 9
   d. 10 − 8 = 2, 8 + 2 = 10   e. 9 − 8 = 1, 8 + 1 = 9   f. 7 − 6 = 1, 6 + 1 = 7

6. a. 3 + 3 = 6 or 6 − 3 = 3. She needs 3 cucumbers.
   b. 7 − 3 = 4. Four ducks are left.
   c. 6 + 2 = 8 or 8 − 6 = 2. She needs 2 dollars.
   d. 6 + 4 = 10 or 10 − 6 = 4. He has 4 pages to read.

7. a. Correct   b. Should be 8 − <u>4</u> = 4   c. Should be <u>6</u> − 4 = 2
   d. Correct   e. Correct   f. Should be 9 − <u>7</u> = 2

8. a. 4 dolls   b. 6 teddy bears   c. 4 other toys

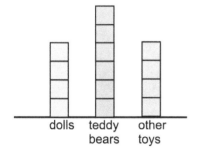

d. 2 more   e. 2 more   f. 10 dolls and teddy bears in all

1. $2 + 7 = 9$; $7 + 2 = 9$; $9 - 2 = 7$; $9 - 7 = 2$

2. a. $10 - 4 = 6$ or $10 - 6 = 4$ b. $5 + 4 = 9$; $9 - 5 = 4$

3. a. $8 - 2 = 6$. Six of them are boys.

    b. $4 + 2 = 6$

    c. $2 + 1 = 3$ robins. $5 - 3 = 2$. Now there are two more sparrows.

4. a. 1, 8, 7, 5   b. 3, 3, 1, 2   c. 10, 2, 0, 3   d. 6, 4, 9, 7

Chapter 3: Place Value Within 0-100

Counting in Groups of 10, p. 112

1.
	a.		b.		c.		d.		e.		f.
ten-groups	ones	ten-groups	ones	ten-groups	ones	ten-groups	ones	ten-groups	ones	ten-groups	ones
2	4	4	2	5	6	5	2	6	3	6	9

4. The teacher needs to check students' work. For example, to make 6 tens and 5 ones in (a), move six full rows of beads and five beads from the seventh row on the abacus.

Naming and Writing Numbers, p. 114

1. a. fifty-six 56
   b. seventy-two 72
   c. twenty-one 21
   d. thirty-one 31
   e. forty-eight 48
   f. thirty-five 35
   g. twenty-three 23
   h. eighty-seven 87
   i. ninety-four 94
   j. sixty-six 66

4. a. twenty-four 2 tens 4 ones = 24
   b. fifty-three 5 tens 3 ones = 53
   c. thirty-nine 3 tens 9 ones = 39
   d. sixty-two 6 tens 2 ones = 62
   e. eighty-eight 8 tens 8 ones = 88
   f. thirty-one 3 tens 1 one = 31
   g. seventy-eight 7 tens 8 ones = 78
   h. ninety-nine 9 tens 9 ones = 99
   i. twenty-two 2 tens 2 ones = 22
   j. twenty-nine 2 tens 9 ones = 29

5. a. 25; twenty-five b. 24; twenty-four c. 36; thirty-six d. 30; thirty
   e. 33; thirty-three f. 40; forty g. 81, eighty-one h. 70; seventy

6. a. 97; ninety-seven b. 64; sixty-four c. 59; fifty-nine d. 100; one hundred

7.

	tens ones		tens ones		tens ones
	2 3		5 0		3 9
	a. twenty-three		b. fifty		c. thirty-nine

The "Teen" Numbers, p. 118

1.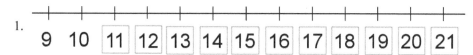
   9 10 11 12 13 14 15 16 17 18 19 20 21

2. a. eleven 11 b. seventeen 17 c. twelve 12 d. sixteen 16
   e. thirteen 13 f. eighteen 18 g. fourteen 14 h. fifteen 15

3. a. 10 + 8 = 18 eighteen
   b. 10 + 1 = 11 eleven
   c. 10 + 4 = 14 fourteen
   d. 10 + 9 = 19 nineteen
   e. 10 + 3 = 13 thirteen
   f. 10 + 2 = 12 twelve
   g. 10 + 7 = 17 seventeen
   h. 10 + 5 = 15 fifteen

The "Teen" Numbers, cont.

4.

1	2	3	4	5	6	7	8	9	10
11	12	13	14	15	16	17	18	19	20
21	22	23	24	25	26	27	28	29	30
31	32	33	34	35	36	37	38	39	40
41	42	43	44	45	46	47	48	49	50
51	52	53	54	55	56	57	58	59	60
61	62	63	64	65	66	67	68	69	70
71	72	73	73	74	75	77	78	79	80
81	82	83	84	85	86	87	88	89	90
91	92	93	94	95	96	97	98	99	100

Building Numbers 11-40, p. 121

1.

a. Eleven	b. Twelve	g. Twenty-one	h. Twenty-two	m. Thirty	n. Thirty-two
10 + 1	10 + 2	20 + 1	20 + 2	30 + 0	30 + 2
tens ones	tens ones	tens ones	tens ones	tens ones	tens ones
1 1	1 2	2 1	2 2	3 0	3 2
c. Fourteen	d. Fifteen	i. Twenty-three	j. Twenty-six	o. Thirty-five	p. Thirty-seven
10 + 4	10 + 5	20 + 3	20 + 6	30 + 5	30 + 7
tens ones	tens ones	tens ones	tens ones	tens ones	tens ones
1 4	1 5	2 3	2 6	3 5	3 7
e. Seventeen	f. Twenty	k. Twenty-seven	l. Twenty-nine	q. Thirty-eight	r. Forty
10 + 7	20 + 0	20 + 7	20 + 9	30 + 8	40 + 0
tens ones	tens ones	tens ones	tens ones	tens ones	tens ones
1 7	2 0	2 7	2 9	3 8	4 0

Building Numbers 41-100, p. 123

1.

a. Forty-one	b. Forty-five	c. Fifty	d. Fifty-seven
40 + 1	40 + 5	50 + 0	50 + 7
tens ones	tens ones	tens ones	tens ones
4 1	4 5	5 0	5 7
e. Seventy-six	f. Eighty-three	g. Ninety-five	h. One hundred
70 + 6	80 + 3	90 + 5	100 + 0 + 0
tens ones	tens ones	tens ones	hundreds tens ones
7 6	8 3	9 5	1 0 0

Building Numbers 41-100, cont.

2.

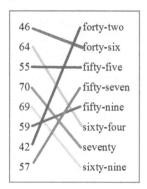 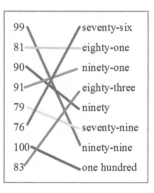

46	forty-two
64	forty-six
55	fifty-five
70	fifty-seven
69	fifty-nine
59	sixty-four
42	seventy
57	sixty-nine

99	seventy-six
81	eighty-one
90	ninety-one
91	eighty-three
79	ninety
76	seventy-nine
100	ninety-nine
83	one hundred

3.

a. 73 = 70 + 3 b. 91 = 90 + 1	c. 45 = 40 + 5 d. 83 = 80 + 3	e. 98 = 90 + 8 f. 64 = 60 + 4

4.

a. 60 + 7 = 67 80 + 0 = 80	b. 4 + 50 = 54 8 + 80 = 88	c. 6 + 80 = 86 0 + 40 = 40	d. 90 + 9 = 99 1 + 60 = 61

5.

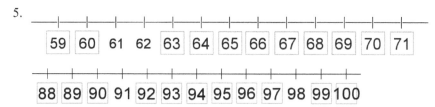

59 60 61 62 63 64 65 66 67 68 69 70 71

88 89 90 91 92 93 94 95 96 97 98 99 100

A 100-Chart, p. 125

1 - 2.

1	2	3	4	5	6	7	8	9	10
11	12	13	14	15	16	17	18	19	20
21	22	23	24	25	26	27	28	29	30
31	32	33	34	35	36	37	38	39	40
41	42	43	44	45	46	47	48	49	50
51	52	53	54	55	56	57	58	59	60
61	62	63	64	65	66	67	68	69	70
71	72	73	74	75	76	77	78	79	80
81	82	83	84	85	86	87	88	89	90
91	92	93	94	95	96	97	98	99	100

4. a. 99, 28   b. 81, 40   c. 85, 69

31

5. The first nine numbers in the column begin with 3, which represents 3 tens, or thirty-something.

6. All of the numbers in that column end in 5, which represents 5 ones.

7. The colored numbers are 28, 53, 76, 82, 100. They are also colored in the 100-chart on the right. Point out to the student that these numbers are right below the given numbers.

8. The underlined numbers are 3, 17, 30, 47, 74, and 89. They are also underlined in the 100-chart on the right. Notice these are directly above the given numbers.

7. and 8. chart

1	2	**3**	4	5	6	7	8	9	10
11	12	**13**	14	15	16	**17**	18	19	20
21	22	23	24	25	26	**27**	28	29	**30**
31	32	33	34	35	36	37	38	39	**40**
41	42	43	44	45	46	**47**	48	49	50
51	52	53	54	55	56	**57**	58	59	60
61	62	63	64	65	66	67	68	69	70
71	72	73	**74**	75	76	77	78	79	80
81	82	83	**84**	85	86	87	88	**89**	90
91	92	93	94	95	96	97	98	**99**	100

9. a. 28, 45   b. 70, 69   c. 14, 27

10.

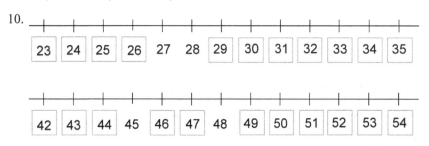

Add and Subtract Whole Tens, p. 127

1. a. 50   b. 80   c. 80   d. 90   e. 100   f. 60   g. 20   h. 20   i. 30   j. 10   k. 50   l. 60

2. a. 60, 80   b. 90, 70   c. 60, 50

3. a. 50, 80   b. 70, 30   c. 10, 60

4. a. 30, 80   b. 50, 40   c. 20, 40

5. a. 50   b. 100

6.

50 50		40 70	
↗ +30 ↘ +10 ↗ −10 ↘ −50		↗ −30 ↘ +10 ↗ +20 ↘ −40	
a. 20 60 0		b. 70 50 30	

Puzzle corner: 99, 101

32

1.

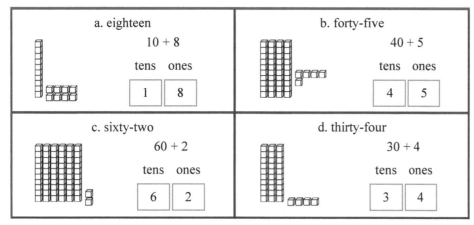

a. eighteen	b. forty-five
10 + 8	40 + 5
tens: 1 ones: 8	tens: 4 ones: 5
c. sixty-two	d. thirty-four
60 + 2	30 + 4
tens: 6 ones: 2	tens: 3 ones: 4

2.

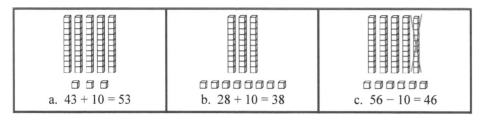

a. 43 + 10 = 53   b. 28 + 10 = 38   c. 56 − 10 = 46

3.

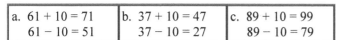

a. 61 + 10 = 71	b. 37 + 10 = 47	c. 89 + 10 = 99
61 − 10 = 51	37 − 10 = 27	89 − 10 = 79

4.

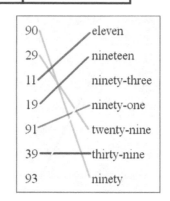

16	twelve
60	fourteen
23	sixteen
32	thirty-two
76	sixty
14	twenty-three
12	seventy-six

90	eleven
29	nineteen
11	ninety-three
19	ninety-one
91	twenty-nine
39	thirty-nine
93	ninety

5. a. 91 = 90 + 1   b. 79 = 70 + 9   c. 58 = 50 + 8

6. a. 26, 82   b. 67, 11   c. 56, 35

7.

| 32 | 33 | 34 | 35 | 36 | 37 | 38 | 39 | 40 | 41 | 42 | 43 | 44 |

| 26 | 27 | 28 | 29 | 30 | 31 | 32 | 33 | 34 | 35 | 36 | 37 | 38 |

Mystery Number: 90

33

Which Number Is Greater?, p. 131

1.

a. 63 is more 3 tens 6 ones — 6 tens 3 ones 36 < 63	b. 54 is more 5 tens 4 ones — 4 tens 5 ones 54 > 45
c. 76 is more 6 tens 7 ones — 7 tens 6 ones 67 < 76	d. 64 is more 4 tens 6 ones — 6 tens 4 ones 46 < 64

Study the above pictures. Do we first check how many TENS the numbers have or how many ONES the numbers have?
Check first how many __tens__ the numbers have.
If the numbers have the same amount of __tens__, then compare the __ones__.

2.

a. 45 < 54	b. 34 > 24	c. 50 < 54
d. 15 < 56	e. 29 < 64	f. 81 < 90
g. 77 > 47	h. 34 < 94	i. 80 > 68

3. a. 81 < 88 b. 95 > 59 c. 99 > 96 d. 85 < 91
e. 90 > 88 f. 49 < 94 g. 90 < 99 h. 100 > 87
i. 50 < 55 j. 22 > 12 k. 98 > 89 l. 24 < 42

4. a. 80 + 2 < 80 + 7 b. 10 + 7 < 70 + 5
c. 60 + 5 > 40 + 5 d. 50 + 7 < 70 + 3
e. 6 + 60 < 6 + 80 f. 5 + 40 = 40 + 5
g. 40 + 5 > 40 + 4 h. 70 + 5 = 5 + 70

5. a. 46 < 67 < 68 b. 33 < 37 < 53 c. 18 < 46 < 48
d. 78 < 80 < 87 e. 46 < 48 < 50 < 84 f. 67 < 76 < 87 < 98

Mystery number: 67

1.

101	102	103	104	105	106	107	108	109	110
111	112	113	114	115	116	117	118	119	120
121	122	123	124	125	126	127	128	129	130

2.

81	82	83	84	85	86	87	88	89	90
91	92	93	94	95	96	97	98	99	100
101	102	103	104	105	106	107	108	109	110

3.

a. one more than 109   <u>110</u>
    one more than 113   <u>114</u>
	b. two more than 116 <u>118</u>
	    two more than 99 <u>101</u>
		c. ten more than 100 <u>110</u>
		    ten more than 102 <u>112</u>

d. one less than 108   <u>107</u>
    one less than 120   <u>119</u>
	e. two less than 110 <u>108</u>
	    two less than 117 <u>115</u>
		f. ten less than 120 <u>110</u>
		    ten less than 107 <u>97</u>

4.

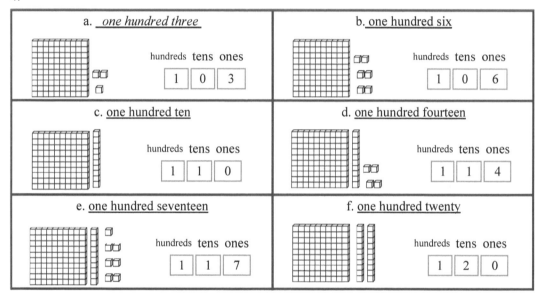

a. *one hundred three* — hundreds 1, tens 0, ones 3

b. one hundred six — hundreds 1, tens 0, ones 6

c. one hundred ten — hundreds 1, tens 1, ones 0

d. one hundred fourteen — hundreds 1, tens 1, ones 4

e. one hundred seventeen — hundreds 1, tens 1, ones 7

f. one hundred twenty — hundreds 1, tens 2, ones 0

5.

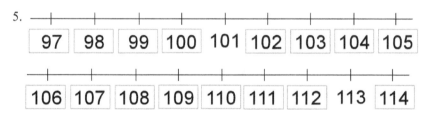

97	98	99	100	101	102	103	104	105

106	107	108	109	110	111	112	113	114

More Practice with Numbers, p. 136

1.

31	32	33	34	35	36	37	38	39	40
41	42	43	44	45	46	47	48	49	50
51	52	53	54	55	56	57	58	59	60
61	62	63	64	65	66	67	68	69	70
71	72	73	74	75	76	77	78	79	80
81	82	83	84	85	86	87	88	89	90
91	92	93	94	95	96	97	98	99	100
101	102	103	104	105	106	107	108	109	110
111	112	113	114	115	116	117	118	119	120

2. a. 77, 97, 55   b. 88, 111, 109   c. 45, 60, 108

3. a. 42   b. 54   c. 38

4. a. 61   b. 26   c. 77   d. 84   e. 75   f. 49

5. a. 20 + 20 + 20 = 60
   b. 30 + 30 + 30 = 90
   c. 20 + 20 + 20 + 20 + 20 = 100

6. a. 88, 89, 90, 91   b. 15, 16, 17, 18
   c. 61, 62, 63, 64   d. 107, 108, 109, 110

Skip-Counting Practice, p. 138

1. a. 0, 2, 4, 6, 8, 10, 12, 14, 16, 18, 20
   b. 50, 52, 54, 56, 58, 60, 62, 64, 66, 68
   c. 1, 3, 5, 7, 9, 11, 13, 15, 17, 19, 21
   d. 27, 29, 31, 33, 35, 37, 39, 41, 43, 45

2.

1	2	3	4	5	6	7	8	9	10
11	12	13	14	15	16	17	18	19	20
21	22	23	24	25	26	27	28	29	30
31	32	33	34	35	36	37	38	39	40
41	42	43	44	45	46	47	48	49	50
51	52	53	54	55	56	57	58	59	60
61	62	63	64	65	66	67	68	69	70
71	72	73	74	75	76	77	78	79	80
81	82	83	84	85	86	87	88	89	90
91	92	93	94	95	96	97	98	99	100

3. a. 0, 4, 8, 12, 16, 20, 24, 28, 32, 36
   b. 52, 56, 60, 64, 68, 72, 76, 80, 84, 88
   c. 1, 5, 9, 13, 17, 21, 25, 29, 33, 37

4.

1	2	3	4	5	6	7	8	9	10
11	12	13	14	15	16	17	18	19	20
21	22	23	24	25	26	27	28	29	30
31	32	33	34	35	36	37	38	39	40
41	42	43	44	45	46	47	48	49	50
51	52	53	54	55	56	57	58	59	60
61	62	63	64	65	66	67	68	69	70
71	72	73	74	75	76	77	78	79	80
81	82	83	84	85	86	87	88	89	90
91	92	93	94	95	96	97	98	99	100

5. a. 0, 10, 20, 30, 40   b. 21, 31, 41, 51, 61
   c. 2, 12, 22, 32, 42   d. 45, 55, 65, 75, 85
   e. 16, 26, 36, 46, 56   f. 59, 69, 79, 89, 99

6. a. 71, 25   b. 44, 92   c. 79, 65

7. a. 0, 5, 10, 15, 20, 25, 30, 35, 40
   b. 1, 6, 11, 16, 21, 26, 31, 36, 41

Mystery Number: 19

Bar Graphs, p. 141

1. a. five   b. ten   c. 7 + 5 = 12 pencils   d. three more

2. a. 30 students   b. 19 students   c. 24 students

3. a. Jane 20   Jerry 35   Jim 15   Hannah 45   Peter 50
   b. Jim read the fewest books.   Peter read the most books.
   c. The two children who read the most books were Hannah and Peter.
      The two children who read the fewest books were Jim and Jane.
   d. Together Jane and Peter read read a total of 70 books. Together Jim and Hannah read read a total of 60 books.

Tally Marks, p. 143

1. a. 6   b. 12   c. 19   d. 23

2.

a. 7 卌 ǁ	b. 14 卌 卌 ǁǁǁ
c. 16 卌 卌 卌 ǀ	d. 32 卌 卌 卌 卌 卌 卌 ǁ
e. 41 卌 卌 卌 卌 卌 卌 卌 卌 ǀ	f. 28 卌 卌 卌 卌 卌 ǁǁǁ

3.

	Tally Marks	Count
Red	卌 ǀ	6
Blue	卌 卌	10
Yellow	卌 卌 ǀ	11

4.

	Tally Marks	Count
Group 1	卌 ǁǁǁ	8
Group 2	卌 卌 ǀ	11

5. The tally marks for Blue, White, and Red cars were in error.

	Tally Marks
Blue	卌 卌 卌 卌 卌
Black	卌 卌
White	卌 卌 卌 卌 卌 卌
Red	卌 卌 卌 卌

1. a. 1 ten 5 ones   _15_   fifteen
   b. 6 tens 7 ones   _67_   sixty-seven
   c. 4 tens 0 ones   _40_   forty
   d. 10 tens 0 ones   _100_   one hundred
   e. 5 tens 1 one   _51_   fifty-one

2.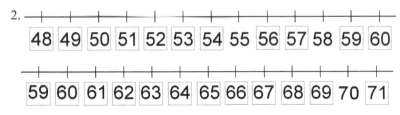

3. a. 87   b. 25   c. 57   d. 80   e. 101

4. 97, 98, 99, 100, 101, 102, 103,
   104, 105, 106, 107, 108, 109

5.

a. $45 = 40 + 5$ $68 = 60 + 8$	b. $25 = 20 + 5$ $54 = 50 + 4$	c. $78 = 70 + 8$ $91 = 90 + 1$

6.

a. $50 + 7 = 57$ $20 + 0 = 20$	b. $8 + 10 = 18$ $9 + 70 = 79$	c. $90 + 6 = 96$ $9 + 60 = 69$

7. a. $17 < 57 < 75$ b. $18 < 41 < 44 < 48$

8. a. 56 $<$ $5 + 60$ b. $20 + 8$ $<$ 33 c. $60 + 5$ $>$ $50 + 6$

   d. 34 $<$ $30 + 6$ e. $4 + 90$ $>$ 49 f. $80 + 2$ $>$ $70 + 9$

9. a. 13, 15, 17, 19, 21, 23, 25, 27, 29
   b. 18, 28, 38, 48, 58, 68, 78, 88, 98
   c. 30, 35, 40, 45, 50, 55, 60, 65, 70

Mystery Number: 54

On the following page is a extra number chart and empty number lines for you to print as needed.

Number Chart _____

Number Lines

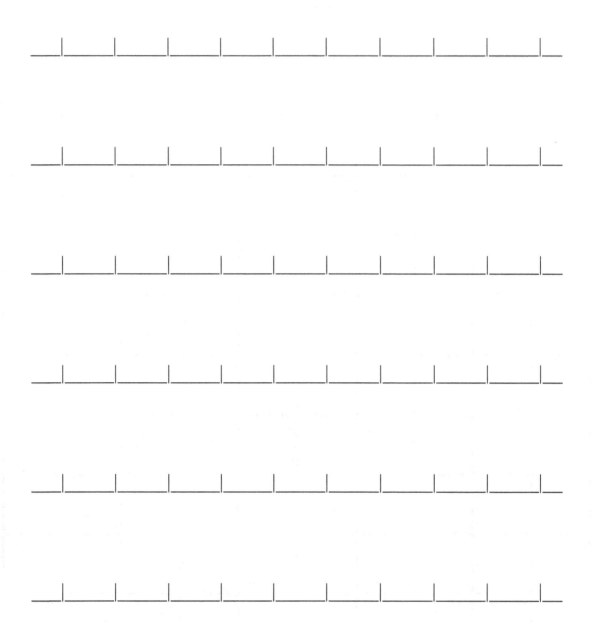

Math Mammoth Grade 1-B
Answer Key

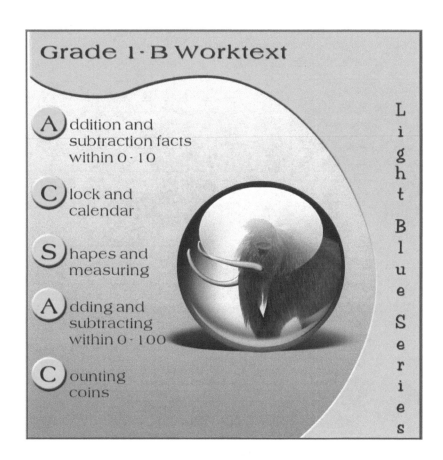

Grade 1·B Worktext

Addition and subtraction facts within 0·10

Clock and calendar

Shapes and measuring

Adding and subtracting within 0·100

Counting coins

Light Blue Series

By Maria Miller

Contents

Chapter 4: Addition and Subtraction Facts

Addition and Subtraction Facts with 4 and 5, p. 10

Facts with 5		$5 + 0 = 5$ $0 + 5 = 5$	$5 - 5 = 0$ $5 - 0 = 5$
		$4 + 1 = 5$ $1 + 4 = 5$	$5 - 4 = 1$ $5 - 1 = 4$
		$3 + 2 = 5$ $2 + 3 = 5$	$5 - 3 = 2$ $5 - 2 = 3$

1. a. 1, 3, 4, 2   b. 3, 4, 1, 2   c. 5, 1, 3, 3   d. 4, 1, 4, 2

2.

$5 - 4$	$2 + 3$	$4 - 4$	$1 + 2$	$4 - 2$	$1 + 3$
$2 + 2$	$3 - 2$	$5 - 0$	$0 + 0$	$5 - 2$	$1 + 1$
$0 + 2$	$5 - 1$	$0 + 1$	$1 + 4$	$0 - 0$	$4 - 1$

3.

$17 - 0 = 17$	$10 + 0 = 10$	$5 - 2 = 3$
$17 - 1 = 16$	$10 + 1 = 11$	$6 - 2 = 4$
$17 - 2 = 15$	$10 + 2 = 12$	$7 - 2 = 5$
$17 - 3 = 14$	$10 + 3 = 13$	$8 - 2 = 6$
$17 - 4 = 13$	$10 + 4 = 14$	$9 - 2 = 7$
$17 - 5 = 12$	$10 + 5 = 15$	$10 - 2 = 8$
$17 - 6 = 11$	$10 + 6 = 16$	$11 - 2 = 9$
$17 - 7 = 10$	$10 + 7 = 17$	$12 - 2 = 10$
$17 - 8 = 9$	$10 + 8 = 18$	$13 - 2 = 11$
$17 - 9 = 8$	$10 + 9 = 19$	$14 - 2 = 12$
$17 - 10 = 7$	$10 + 10 = 20$	$15 - 2 = 13$
$17 - 11 = 6$	$10 + 11 = 21$	$16 - 2 = 14$
$17 - 12 = 5$	$10 + 12 = 22$	$17 - 2 = 15$
etc.	etc.	etc.

1.

6, 0, 6
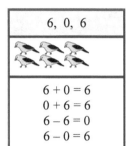
6 + 0 = 6 0 + 6 = 6 6 − 6 = 0 6 − 0 = 6

5, 1, 6
5 + 1 = 6 1 + 5 = 6 6 − 5 = 1 6 − 1 = 5

4, 2, 6
4 + 2 = 6 2 + 4 = 6 6 − 4 = 2 6 − 2 = 4

3, 3, 6
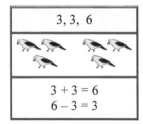
3 + 3 = 6 6 − 3 = 3

2.

0 + 6 = 6	or	6 + 0 = 6
1 + 5 = 6	or	5 + 1 = 6
2 + 4 = 6	or	4 + 2 = 6
3 + 3 = 6		

3. a.
$$\begin{array}{r} 6 \\ -5 \\ \hline 1 \end{array}$$
b.
$$\begin{array}{r} 6 \\ -4 \\ \hline 2 \end{array}$$
c.
$$\begin{array}{r} 6 \\ -6 \\ \hline 0 \end{array}$$
d.
$$\begin{array}{r} 6 \\ -2 \\ \hline 4 \end{array}$$
e.
$$\begin{array}{r} 6 \\ -1 \\ \hline 5 \end{array}$$
f.
$$\begin{array}{r} 6 \\ -3 \\ \hline 3 \end{array}$$

5. a. 2, 3   b. 1, 6   c. 4, 5   d. 4, 1

6.

a. 2 + △3 = 5
  5 − 2 = △3
	b. 1 + △5 = 6
	  6 − 1 = △5
		c. 4 + △1 = 5
		  5 − 4 = △1

d. 3 + △5 = 8
  8 − 3 = △5
	e. 5 + △5 = 10
	  10 − 5 = △5
		f. 2 + △5 = 7
		  7 − 2 = △5

7.

a. 1 + △5 = 6
  △5 + 1 = 6
  6 − △5 = 1
  6 − 1 = △5
	b. 2 + 7 = △9
	  7 + 2 = △9
	  △9 − 2 = 7
	  △9 − 7 = 2

8. a. 4 + 3 = 7. They have seven kittens altogether.
   4 − 3 = 1 OR 3 + 1 = 4. The black cat has one more than the white one.

   b. 10 − 2 = 8 OR 2 + 8 = 10. He has lost eight crayons.

   c. 10 + 2 = 12. Mother had 12 clothespins. 10 − 2 = 8 OR 8 + 2 = 10.
   Mother had eight more in the first container.

   d. 8 − 2 = 6 OR 2 + 6 = 8. She needs six more eggs.
   2 + 3 = 5. Jill and the neighbor have five eggs in total.
   8 − 5 = 3 OR 5 + 3 = 8. So, she will need three more.

1.

7, 0, 7	6, 1, 7	5, 2, 7
$7 + 0 = 7$ $0 + 7 = 7$ $7 - 0 = 7$ $7 - 7 = 0$	$6 + 1 = 7$ $1 + 6 = 7$ $7 - 6 = 1$ $7 - 1 = 6$	$5 + 2 = 7$ $2 + 5 = 7$ $7 - 5 = 2$ $7 - 2 = 5$

4, 3, 7
$4 + 3 = 7$ $3 + 4 = 7$ $7 - 4 = 3$ $7 - 3 = 4$

2.

$0 + 7 = 7$	or	$7 + 0 = 7$
$1 + 6 = 7$	or	$6 + 1 = 7$
$2 + 5 = 7$	or	$5 + 2 = 7$
$3 + 4 = 7$	or	$4 + 3 = 7$

4.
a. $\begin{array}{r} 7 \\ -5 \\ \hline 2 \end{array}$
b. $\begin{array}{r} 7 \\ -4 \\ \hline 3 \end{array}$
c. $\begin{array}{r} 7 \\ -6 \\ \hline 1 \end{array}$
d. $\begin{array}{r} 7 \\ -2 \\ \hline 5 \end{array}$
e. $\begin{array}{r} 7 \\ -1 \\ \hline 6 \end{array}$
f. $\begin{array}{r} 7 \\ -3 \\ \hline 4 \end{array}$

5.

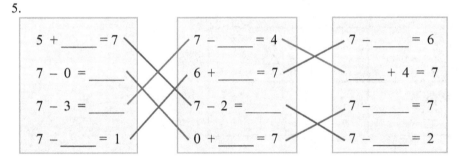

$5 + \_\_\_ = 7$

$7 - 0 = \_\_\_$

$7 - 3 = \_\_\_$

$7 - \_\_\_ = 1$

$7 - \_\_\_ = 4$

$6 + \_\_\_ = 7$

$7 - 2 = \_\_\_$

$0 + \_\_\_ = 7$

$7 - \_\_\_ = 6$

$\_\_\_ + 4 = 7$

$7 - \_\_\_ = 7$

$7 - \_\_\_ = 2$

6. a. Jeremy has two more pencils than Luis. $6 - 4 = 2$ or $4 + \_\_\_ = 6$.
   Altogether they have 10 pencils. $6 + 4 = 10$.
   b. She found eight socks. $2 + 5 + 1 = 8$.

Puzzle corner:

−	12	11	10	9	8	7	6	5	4	3
1	11	10	9	8	7	6	5	4	3	2
2	10	9	8	7	6	5	4	3	2	1

1.

8, _0_, 8
8 + 0 = 8
0 + 8 = 8
8 − 0 = 8
8 − 8 = 0

_7_, 1, 8
7 + 1 = 8
1 + 7 = 8
8 − 1 = 7
8 − 7 = 1

6, 2, 8
6 + 2 = 8
2 + 6 = 8
8 − 2 = 6
8 − 6 = 2

5, 3, 8
5 + 3 = 8
3 + 5 = 8
8 − 5 = 3
8 − 3 = 5

4, 4, 8
4 + 4 = 8
8 − 4 = 4

3.

0 + _8_ = 8	or	_8_ + 0 = 8	
1 + _7_ = 8	or	_7_ + 1 = 8	
2 + _6_ = 8	or	_6_ + _2_ = 8	4 + _4_ = 8
3 + _5_ = 8	or	_5_ + _3_ = 8	

4. a. 5, 7, 6   b. 3, 7, 6   c. 4, 2, 1   d. 1, 8, 4

5.

a. 5 − 2 4	b. 7 − 4 5	c. 8 − 1 7	d. 6 − 3 2
↓ ↓	↓ ↓	↓ ↓	↓ ↓
3 < 4	3 < 5	7 = 7	3 > 2

6.

a. $\quad$ 5 – 2 $\qquad$ 4 – 2 $\downarrow \qquad \downarrow$ $\boxed{3} > \boxed{2}$	b. $\quad$ 8 – 1 $\qquad$ 7 – 1 $\downarrow \qquad \downarrow$ $\boxed{7} > \boxed{6}$

a. 5 – 2 4 – 2	b. 8 – 1 7 – 1	c. 8 – 6 8 – 5
↓ ↓ 3 > 2	↓ ↓ 7 > 6	↓ ↓ 2 < 3
d. 6 + 2 7 + 2 ↓ ↓ 8 < 9	e. 7 – 1 7 – 2 ↓ ↓ 6 > 5	f. 4 + 4 7 – 5 ↓ ↓ 8 > 2
g. 1 – 1 3 – 2 ↓ ↓ 0 < 1	h. 3 + 10 10 ↓ ↓ 13 > 10	i. 7 4 + 2 ↓ ↓ 7 > 6
j. 8 – 1 4 – 2 ↓ ↓ 7 > 2	k. 7 – 2 6 – 1 ↓ ↓ 5 = 5	l. 9 – 0 7 + 2 ↓ ↓ 9 = 9

7.

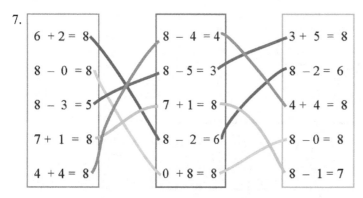

6 + 2 = 8	8 – 4 = 4	3 + 5 = 8
8 – 0 = 8	8 – 5 = 3	8 – 2 = 6
8 – 3 = 5	7 + 1 = 8	4 + 4 = 8
7 + 1 = 8	8 – 2 = 6	8 – 0 = 8
4 + 4 = 8	0 + 8 = 8	8 – 1 = 7

8. a. 7 – 4 = 3. Bill has three more cars than Ed.
$\quad$ 10 – 4 = 6. Jack has six more cars than Ed.
$\quad$ 10 – 7 = 3. Jack has three more cars than Bill.

$\quad$ b. 5 + 3 = 8. Both things together cost $8, so she cannot buy both. She has $7.
$\quad\quad$ 7 + 1 = 8. She would need $1 more.

1.

9, 0, 9
☆☆☆☆☆ ☆☆☆☆
9 + 0 = 9
0 + 9 = 9
9 − 0 = 9
9 − 9 = 0

8, 1, 9
☆☆☆☆ ☆ ☆☆☆☆
8 + 1 = 9
1 + 8 = 9
9 − 1 = 8
9 − 8 = 1

7, 2, 9
☆☆☆☆ ☆ ☆☆☆ ☆
7 + 2 = 9
2 + 7 = 9
9 − 7 = 2
9 − 2 = 7

6, 3, 9
☆☆☆ ☆☆ ☆☆☆ ☆
6 + 3 = 9
3 + 6 = 9
9 − 6 = 3
9 − 3 = 6

5, 4, 9
☆☆☆ ☆☆ ☆☆ ☆☆
5 + 4 = 9
4 + 5 = 9
9 − 5 = 4
9 − 4 = 5

3.

0 + _9_ = 9	or _9_ + 0 = 9
1 + _8_ = 9	or _8_ + 1 = 9
2 + _7_ = 9	or _7_ + _2_ = 9
3 + _6_ = 9	or _6_ + _3_ = 9
4 + _5_ = 9	or _5_ + _4_ = 9

4. a. 4, 6, 3, 1   b. 7, 8, 2, 1   c. 8, 6, 4, 2   d. 8, 9, 7, 5

5.

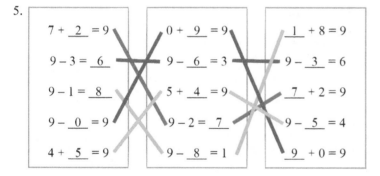

7 + _2_ = 9	0 + _9_ = 9	_1_ + 8 = 9
9 − 3 = _6_	9 − _6_ = 3	9 − _3_ = 6
9 − 1 = _8_	5 + _4_ = 9	_7_ + 2 = 9
9 − _0_ = 9	9 − 2 = _7_	9 − _5_ = 4
4 + _5_ = 9	9 − _8_ = 1	_9_ + 0 = 9

6. a.   8 > 10 − 3   b.   9 < 9 + 3   c.   8 − 6 < 6 + 3

d. 6 + 2 < 8 + 2   e.   10 − 1 < 10   f.   8 − 4 > 8 − 5

g. 5 − 2 > 4 − 2   h.   8 + 0 = 8 − 0   i.   9 − 1 < 9 + 1

7. a.   9   b.   9   c.   9   d.   8   e.   9   f.   8
      − 5   − 4   − 6   − 2   − 2   − 3
      ———   ———   ———   ———   ———   ———
      4   5   3   6   7   5

8.

$9-3$	$4+6$	$9-0$	$4+1$	$8-1$
$2+5$	$9-5$	$4+4$	$4-2$	$5+1$
$9-2$	$3+7$	$10-2$	$10+0$	$7-1$
$4+2$	$7-3$	$6+3$	$3-1$	$3+3$
$6-0$	$1+1$	$8-0$	$3+2$	$10-4$
$3+4$	$8-3$	$2+7$	$7-6$	$7+0$
$1+6$	$2+8$	$10-1$	$2+2$	$7-0$

($9-3$, $2+5$, $9-2$, $4+2$, $6-0$, $3+4$, $1+6$, $8-1$, $5+1$, $7-1$, $3+3$, $10-4$, $7+0$, $7-0$ are colored blue, $9-0$, $4+4$, $10-2$, $6+3$, $8-0$, $2+7$, $10-1$ are colored red, and the rest are colored yellow.)

Addition and Subtraction Facts with 10, p. 24

1.

10, 0, 10
$10+0=10$
$0+10=10$
$10-10=0$
$10-0=10$

9, 1, 10
$9+1=10$
$1+9=10$
$10-9=1$
$10-1=9$

8, 2, 10
$8+2=10$
$2+8=10$
$10-2=8$
$10-8=2$

7, 3, 10
$7+3=10$
$3+7=10$
$10-7=3$
$10-3=7$

6, 4, 10
$6+4=10$
$4+6=10$
$10-6=4$
$10-4=6$

5, 5, 10
$5+5=10$
$10-5=5$

3.

$0+\underline{10}=10$	or	$\underline{10}+0=10$
$1+\underline{9}=10$	or	$\underline{9}+1=10$
$2+\underline{8}=10$	or	$\underline{8}+\underline{2}=10$
$3+\underline{7}=10$	or	$\underline{7}+\underline{3}=10$
$4+\underline{6}=10$	or	$\underline{6}+\underline{4}=10$
$5+\underline{5}=10$		

4. a. 7, 4, 8   b. 10, 3, 1   c. 3, 8, 6

Addition and Subtraction Facts with 10, cont.

5.

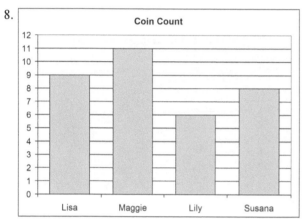

6. a. 2 + 3 = 5. Elisa has five coins. Sarah still has more coins. 6 − 5 = 1. The difference is one coin.
   b. 6 + 1 = 7. Dad got a total of seven boxes. Then he gave one away. 7 − 1 = 6. So he ended up with six boxes.
   c. 6 + 3 = 9. We have a total of nine dollars, so yes, we can buy the meal. 9 − 8 = 1. We will have one dollar left over.

7. a. 5, 3, 8   b. 7, 6, 9   c. 4, 3, 1

8.

```
                    Coin Count
12
11            ┌────┐
10            │    │
 9  ┌────┐    │    │
 8  │    │    │    │              ┌────┐
 7  │    │    │    │              │    │
 6  │    │    │    │   ┌────┐     │    │
 5  │    │    │    │   │    │     │    │
 4  │    │    │    │   │    │     │    │
 3  │    │    │    │   │    │     │    │
 2  │    │    │    │   │    │     │    │
 1  │    │    │    │   │    │     │    │
 0  └────┘    └────┘   └────┘     └────┘
     Lisa     Maggie    Lily     Susana
```

9. Answers will vary; check the student's questions and answers. For example: How many more coins does
   Lisa have than Maggie?

Puzzle Corner.
Use "guess and check" to find the answers.
a. The square is 6, the triangle is 4. 6 + 4 = 10, 6 − 4 = 2
b. The square is 8, the triangle is 2. 8 + 2 = 10, 8 − 2 = 6
c. The square is 5, and the triangle is 5.   5 + 5 = 10, 5 − 5 = 0

Subtracting More Than One Number, p. 28

1. a. 3, 1, 4   b. 5, 3, 2   c. 2, 2, 5

2. a. She has six cookies left. 10 − 2 − 2 = 6   b. There are three birds left. 7 − 3 − 1 = 3
   c. Now there are three cars. 8 − 3 − 2 = 3   d. Jack now has $8. $5 + $2 + $1 = $8

3. a. 2, 1   b. 1, 2   c. 1, 3

4. a. 5, 4   b. 1, 0   c. 2, 1

5.

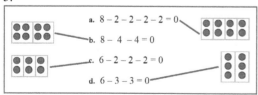

Puzzle corner. 9 − 3 − 2 − 1 = 3   10 − 1 − 2 − 1 = 6   8 − 4 − 1 − 2 = 1

1.

a.	b.	c.	d.
$0 + 8 = 8$	$3 + 4 = 7$	$6 - 4 = 2$	$7 - 5 = 2$
$3 + 5 = 8$	$5 + 2 = 7$	$6 - 1 = 5$	$8 - 5 = 3$
$2 + 6 = 8$	$1 + 6 = 7$	$6 - 3 = 3$	$6 - 5 = 1$
$6 + 2 = 8$	$6 + 1 = 7$	$6 - 2 = 4$	$8 - 4 = 4$
$5 + 3 = 8$	$2 + 5 = 7$	$6 - 5 = 1$	$7 - 3 = 4$

2.

a. $8 - 2 \;\square\; 7 - 3$	b. $10 - 7 \;\square\; 9 - 6$	c. $7 - 6 \;\square\; 4 - 2$
$\downarrow \qquad \downarrow$	$\downarrow \qquad \downarrow$	$\downarrow \qquad \downarrow$
$6 \;>\; 4$	$3 \;=\; 3$	$1 \;<\; 2$
d. $\quad 4 + 2 \;>\; 9 - 8$	e. $\quad 10 - 4 \;>\; 7 - 4$	f. $\quad 3 + 4 \;>\; 7 - 1$

3. a. Luisa had 5 more counters. $9 - 4 = 5$ or $4 + 5 = 9$.
   b. Luisa had three more counters. $8 - 5 = 3$ or $5 + 3 = 8$.

4.

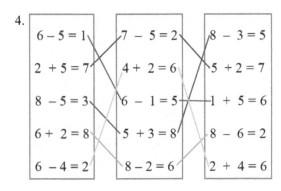

5.

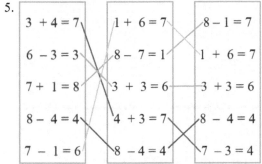

Puzzle Corner.
Answers will vary.
Here is one
possibility.

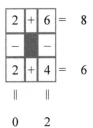

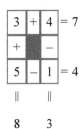

1.

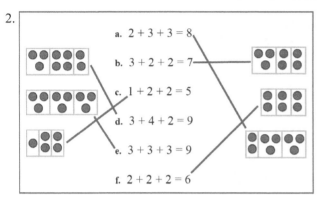

a.
4 + _5_ = 9
1 + _8_ = 9
6 + _3_ = 9
2 + _7_ = 9

b.
5 + _5_ = 10
2 + _8_ = 10
3 + _7_ = 10
4 + _6_ = 10

c.
10 − _9_ = 1
10 − _3_ = 7
10 − _5_ = 5
10 − _2_ = 8

d.
9 − _7_ = 2
9 − _3_ = 6
9 − _1_ = 8
9 − _4_ = 5

2.

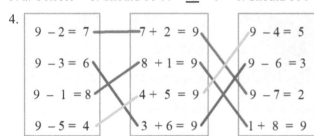

a. 2 + 3 + 3 = 8
b. 3 + 2 + 2 = 7
c. 1 + 2 + 2 = 5
d. 3 + 4 + 2 = 9
e. 3 + 3 + 3 = 9
f. 2 + 2 + 2 = 6

3. a. Correct   b. Should be 10 − _4_ = 6   c. Should be 9 − 4 = _5_   d. Correct   e. Should be 7 − _4_ = 3   f. Correct

4.

9 − 2 = 7	7 + 2 = 9	9 − 4 = 5
9 − 3 = 6	8 + 1 = 9	9 − 6 = 3
9 − 1 = 8	4 + 5 = 9	9 − 7 = 2
9 − 5 = 4	3 + 6 = 9	1 + 8 = 9

5. a. Answers may vary. Please check the student's work. In part (a) one number 9 will be left unpaired.
   In part (b) a number 6 will be left unpaired.

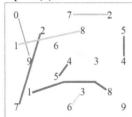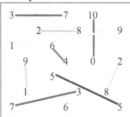

6.

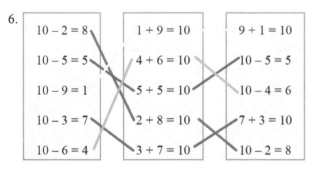

10 − 2 = 8	1 + 9 = 10	9 + 1 = 10
10 − 5 = 5	4 + 6 = 10	10 − 5 = 5
10 − 9 = 1	5 + 5 = 10	10 − 4 = 6
10 − 3 = 7	2 + 8 = 10	7 + 3 = 10
10 − 6 = 4	3 + 7 = 10	10 − 2 = 8

7. a. 7 − 2 = 5. Ken has five more than Millie.
   b. 3 + 4 + 3 = 10. Mike has ten cars.
   c. 4 + 4 = 8. There were eight birds. 8 − 5 = 3. Later, there were three birds.
   d. 4 + 6 = 10 or 10 − 4 = 6. Six crayons are missing.
   e. 10 − 2 = 8. There are eight pieces left.

Chapter 5: Time

Whole and Half Hours, p. 38

1. a. 2 o'clock   b. 11 o'clock   c. 6 o'clock   d. 9 o'clock

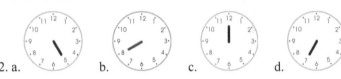

2. a.   b.   c.   d.

3. a. half past 1   b. half past 8   c. half past 12   d. half past 5.

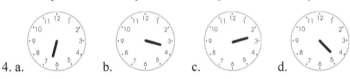

4. a.   b.   c.   d.

5. a. 5 o'clock   b. half past 12   c. half past 7   d. half past 2
   e. 4 o'clock   f. 7 o'clock   g. 12 o'clock   h. half past 1

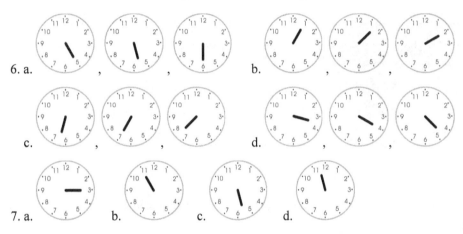

6. a.   ,   ,   b.   ,   ,

c.   ,   ,   d.   ,   ,

7. a.   b.   c.   d.

   a. half past 3   b. half past 11   c. 6 o'clock   d. 12 o'clock

8. a. 5 o'clock, 6 o'clock   b. 12 o'clock, 1 o'clock   c. half past 3, half past 4   d. half past 9, half past 10

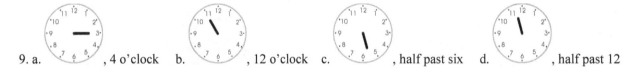

9. a.   , 4 o'clock   b.   , 12 o'clock   c.   , half past six   d.   , half past 12

Minutes and Half Hours, p. 42

1. a. Half past 6   b. 5 o'clock   c. Half past 2   d. Half past 10

2. a. 6 o'clock, 6:00   b. Half past 6, 6:30   c. 2 o'clock, 2:00   d. Half past 9, 9:30
   e. Half past 4, 4:30   f. Half past 12, 12:30   g. 11 o'clock, 11:00   h. 5 o'clock, 5:00

3. a. 3:30, 4:00   b. 7:00, 7:30   c. 9:30, 10:00   d. 3:00, 3:30   e. 12:00, 12:30

4. a. 3:00   b. 6:30   c. 12:30   d. 1:00   e. 9:30

5. a. 3 hours   b. 1 hour   c. 1/2 an hour   d. 2 hours   e. 5 hours   f. 1/2 an hour

6. a. 9:35   b. 6:10   c. 3:25   d. 11:35   e. 1:15   f. 11:50   g. 7:05   h. 12:45

7. a. 9:10   b. 12:10   c. 7:05   d. 5:20   e. 7:55   f. 1:50   g. 8:45   h. 7:40

Time Order, p. 46

1. The answers will vary.

2. a. <u>Yesterday</u> Paul studied for the test.
   <u>Tomorrow</u> Paul will know the results.
   <u>Today</u> Paul has a math test.

   b. <u>Today</u> Mom makes a birthday cake!
   <u>Tomorrow</u> we eat the leftovers.
   <u>Yesterday</u> Mom bought eggs, flour, sugar, and fruit.

3.

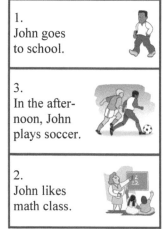

3. Jane's ankle is hurt.	2. Colored pencils!	1. John goes to school.
1. Jane rides her bike.	1. Rick gets a gift.	3. In the after-noon, John plays soccer.
2. Jane crashes into a tree.	3. Rick draws an airplane.	2. John likes math class.

4. a. Prepare a meal.   b. Read a book.   c. Clean the house.
   d. A soccer game.   e. Go shopping for food.   f. A game of cards.

AM and PM, p. 48

1. Answers may vary.

2. These are example solutions. The actual sunrise and sunset times in your location may vary.

a.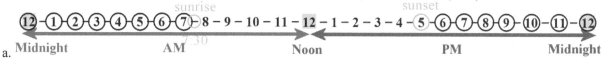

b.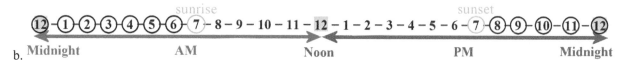

c.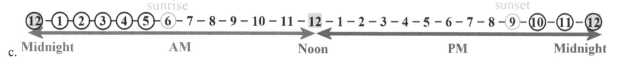

3. Here are some example answers:
   a. sleeping   b. schoolwork   c. eating lunch   d. playing   e. eating supper   f. going to bed   g. sleeping

4. a. It is before noon, 5:00 AM   b. No, 8:00 PM   c. Yes, 8:00 AM   d. Yes, 11:00 AM
   e. Yes, 2:30 AM   f. Yes, 12:30 AM   g. No, 12:30 PM   h. No, 10:30 PM

5. a.   , 7:00 AM   b.   , 12:00 PM   c.   , 6:00 PM
   Answers will vary according to the student's schedule.

1. a.- f. g., h, and i. answers will vary.

January	February	March
Su Mo Tu We Th Fr Sa	Su Mo Tu We Th Fr Sa	Su Mo Tu We Th Fr Sa
1 2 3	1 2 3 4 5 6 7	1 2 3 4 5 6 7
4 5 6 7 8 9 10	d. 8 9 10 11 12 13 14	a. 8 9 10 11 12 13 14
11 12 13 14 15 16 17	15 16 17 18 19 20 21	15 16 17 18 19 20 21
18 19 20 21 22 23 24	22 23 24 25 26 27 28	22 23 24 25 26 27 28
25 26 27 28 29 30 31		29 30 31

April	May	June
Su Mo Tu We Th Fr Sa	Su Mo Tu We Th Fr Sa	Su Mo Tu We Th Fr Sa
1 2 3 4	1 2	1 2 3 4 5 6
5 6 7 8 9 10 11	3 4 5 6 7 8 9	7 8 9 10 11 12 13
12 13 14 15 16 17 18	10 11 12 13 14 15 16	14 15 16 17 18 19 20
19 20 21 22 23 24 25	17 18 19 20 21 22 23	21 22 23 24 25 26 27
26 27 28 29 30	24 25 26 27 28 29 30	28 29 30
	31	

July	August	September
Su Mo Tu We Th Fr Sa	Su Mo Tu We Th Fr Sa	Su Mo Tu We Th Fr Sa
1 2 3 4	1	1 2 3 4 5
5 6 7 8 9 10 11	2 3 4 5 6 7 8	6 7 8 9 10 11 12
12 13 14 15 16 17 18	9 10 11 12 13 14 15	13 14 15 16 17 18 19
19 20 21 22 23 24 25	16 17 18 19 20 21 22	20 21 22 23 24 25 26
26 27 28 29 30 31	23 24 25 26 27 28 29	27 28 29 30
	30 31	

October	November	December
Su Mo Tu We Th Fr Sa	Su Mo Tu We Th Fr Sa	Su Mo Tu We Th Fr Sa
1 2 3	1 2 3 4 5 6 7	1 2 3 4 5
4 5 6 7 8 9 10	8 9 10 e.11 12 13 14	6 7 8 9 10 11 12
11 12 13 14 15 16 17	15 16 17 b.18 19 20 21	13 14 15 16 17 18 19
18 19 20 21 22 23 24	22 23 24 25 26 27 28	20 21 22 23 24 25 26
25 26 27 28 29 f.30 31	29 30	27 28 c.29 30 31

2. Su = Sunday,   Mo = Monday,   Tu = Tuesday,   We = Wednesday,   Th = Thursday,   Fr = Friday,   Sa = Saturday

3. a. August   <u>September</u>   October
   b. March   <u>April</u>   May
   c. June   <u>July</u>   August
   d. January   <u>February</u>   March
   e. October   <u>November</u>   December

4. a. January 4 is <u>Sunday.</u>   b. May 14 is <u>Thursday</u>
   c. July 23 is <u>Thursday.</u>   d. March 10 is <u>Tuesday.</u>
   e. Today (answers will vary)   f. Your birthday (answers will vary)

5. June <u>1</u>, June <u>8</u> , June <u>15</u>, June <u>22</u> , June <u>29</u>

6. April <u>2</u>, April <u>9</u>, April <u>16</u> , April <u>23</u> , April <u>30</u>

7. January, February, March, April;
   May, June, July, August;
   September, October, November, December.

1. a. half past 7   b. 2 o'clock   c. half past 12   d. half past 3

2. a. 9:00   b. 9:30   c. 4:30   d. 12:00

3. a. 5:30, 6:00   b. 11:00, 11:30   c. 12:30, 1:00   d. 2:00, 2:30   e. 6:00, 6:30

4. a. AM   b. PM   c. AM   d. PM

Chapter 6: Shapes and Measuring

Basic Shapes, p. 57

1.
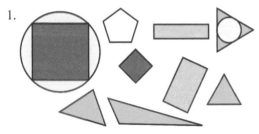

2. a. 3   b. 5   c. 5   d. 0   e. 6
   f. 4   g. 4   h. 0   i. 4   j. 7

3. a. R   b. 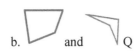 and Q

   c. C   d. It is an oval.

4. a. You get a triangle.
   b. You get a triangle again, unless you draw the three dots so that they are "perfectly aligned," so that joining them you just get a line.

5.

a. The new shapes have _4_ sides, and _4_ corners.
They are _squares_ .
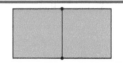

b. The new shapes have _3_ sides, and _3_ corners.
They are _triangles._

c. The new shapes have _4_ sides, and _4_ corners.
They are _quadrilaterals_

d. The new shapes have _3_ sides, and _3_ corners.
They are _triangles._

e. The new shapes have _3_ sides, and _3_ corners.
They are _triangles._

Puzzle corner:

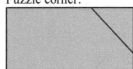

1.

2.

3. or the two rectangles side-by side

4. or this combination in other positions

5.

6.

7. or

8. or

9. a. b.

c. One possibility:

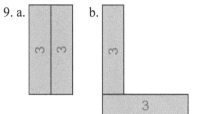

10.

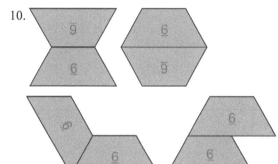

11.

12.

13. Answers will vary. For example:

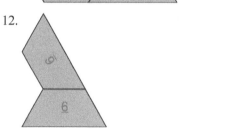

14. Answers will vary. For example:

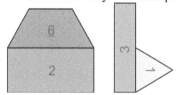

61

1.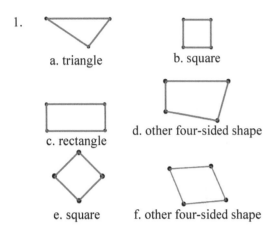

   a. triangle   b. square

   c. rectangle   d. other four-sided shape

   e. square   f. other four-sided shape

2. Answers will vary.

3. What kind of shapes do you get now?
   <u>triangles</u>
   How many parts does each four-sided
   shape have now? <u>4</u>
   What kind of shapes are these parts?
   <u>triangles</u>

   a.   b.

   c.   d.

4. Answers will vary since the student can choose the colors.
   For example:

   Triangles are <u>blue.</u>   Circles are <u>yellow</u>.
   Squares are <u>purple</u>.   Rectangles are <u>green</u>.
   Other four-sided shapes are <u>red</u>.

   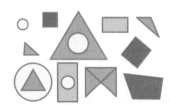

1. The colors that the student uses will vary.
   As long as the new shapes aren't the same
   color, it doesn't matter what color they are.

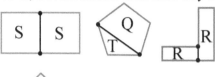

2.

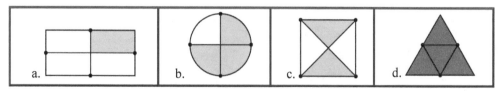

3.

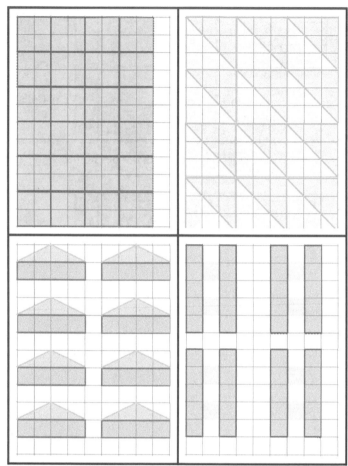

4. Answers will vary.

Halves and Quarters, p. 69

1.

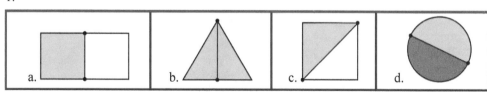

2.

3.

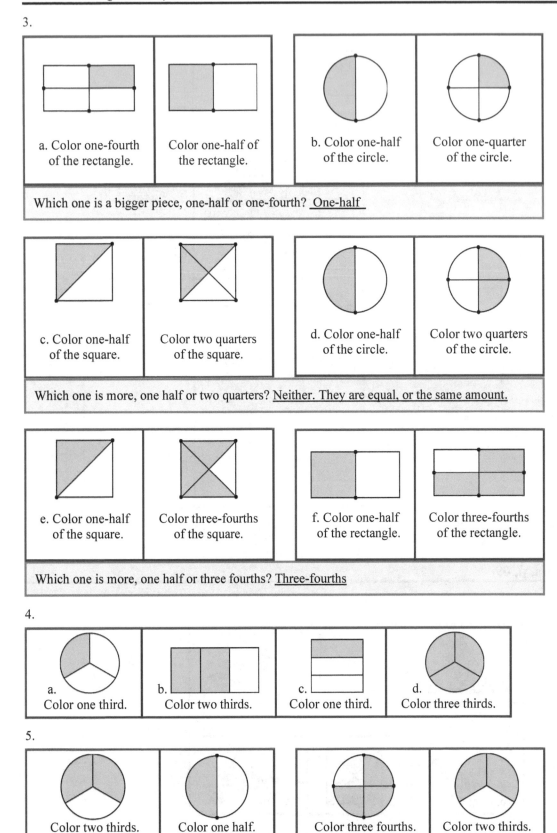

a. Color one-fourth of the rectangle.

Color one-half of the rectangle.

b. Color one-half of the circle.

Color one-quarter of the circle.

Which one is a bigger piece, one-half or one-fourth? <u>One-half</u>

c. Color one-half of the square.

Color two quarters of the square.

d. Color one-half of the circle.

Color two quarters of the circle.

Which one is more, one half or two quarters? <u>Neither. They are equal, or the same amount.</u>

e. Color one-half of the square.

Color three-fourths of the square.

f. Color one-half of the rectangle.

Color three-fourths of the rectangle.

Which one is more, one half or three fourths? <u>Three-fourths</u>

4.

a. Color one third.

b. Color two thirds.

c. Color one third.

d. Color three thirds.

5.

Color two thirds.

Color one half.

Color three fourths.

Color two thirds.

a. Which is more, two thirds or one half? <u>two thirds</u>

b. Which is more, three fourths or two thirds? <u>three fourths</u>

Halves and Quarters, cont.

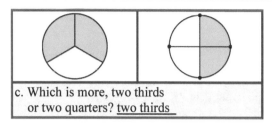 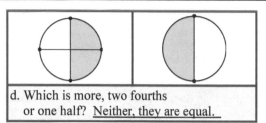

c. Which is more, two thirds
   or two quarters? <u>two thirds</u>

d. Which is more, two fourths
   or one half? <u>Neither, they are equal.</u>

6.

a. Which is more, one half or one third? <u>one half</u>
b. Which is more, one fourth or one third? <u>one third</u>

7.

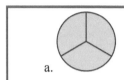   a.

The whole pie is
<u>3</u> <u>thirds</u> .

   b.

The whole pie is
<u>2</u> <u>halves</u> .

   c.

The whole pie is
<u>4</u>   <u>quarters/fourths</u>.

8.

a. 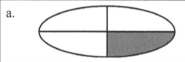   <u>1</u>   <u>fourth</u> of
the _____<u>oval</u>_____ is colored.

b. <u>1</u> _____<u>third</u>_____ of
the _____<u>hexagon</u>_____ is colored.

c.   <u>1</u> <u>half</u> of
the _____<u>trapezoid</u>_____ is colored.

d. <u>2</u> <u>thirds</u> of
the <u>rectangle</u> are colored.

e.   <u>3</u> <u>quarters/fourths</u> of
the <u>circle</u> are colored.

f. <u>2</u> <u>quarters/fourths</u> of
the <u>triangle</u> are colored.

65

Measuring Length, p. 73

1. a. Please check the student's answers.
   b. Please check the student's answers.

2. The desk was about 12 baby-shoes wide.

3. Ryan's room was 81 baby-shoes wide.
   Since baby-shoes are smaller than daddy-shoes, we know the number will be larger than 27.

4. Answers will vary.

5. pencil a. __5__ crayons long; __10__ paperclips long
   pencil b. __6__ crayons long; __12__ paperclips long
   pencil c. __4__ crayons long; __8__ paperclips long
   pencil d. __3__ crayons long; __6__ paperclips long

6. a. The pen is longer than the pencil.
   b. The flashlight is longer than the celery.
   c. The blue (second) car is longer than the pink car.
   d. The toothbrush is longer than the pencil.

7. a.   b.

Exploring Measuring, p. 77

1. Answers will vary. Check the student's answers.

2. a. sheep 1, car 2, dinosaur 3   b. crayon 1, carrot 2, celery 3   c. paperclip 1, toothbrush 2, flashlight 3

3. Answers will vary.

4. Answers will vary.

5. Answers will vary.

6. The bucket held 53 smaller drinking glasses. It will hold more small glassfuls than large ones.

Measuring Lines in Inches, p. 79

1. a. 2 in.   b. 4 in.   c. 3 in.   d. 5 in.

2. 1 in., 5 in., 3 in., 6 in., 5 in.

3.

5. Check the student's lines.

6.

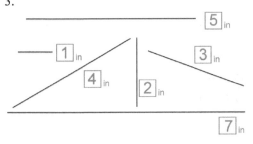

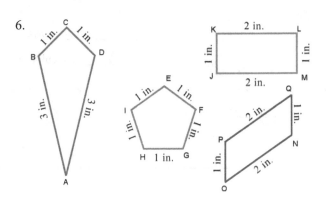

4. Triangle ABC: AB is 2 in., BC is 3 in., AC is 4 in.
   Triangle DEF: DF is 5 in., DE is 4 in., EF is 3 in.

Measuring Lines in Centimeters, p. 82

1. a. 5 cm   b. 3 cm   c. 9 cm   d. 10 cm   e. 12 cm

2.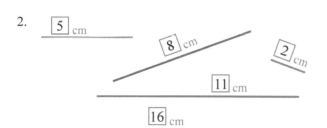

3.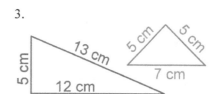

4. Check the student's lines.

5. Answers will vary.

Three-Dimensional Shapes p. 84

1. a. box   b. cube   c. box   d. cube   e. box   f. box   g. cube   h. box

2. Answers will vary. Please check the student's work.

3. Answers will vary. Please check the student's work.

4. a. ball   b. cylinder   c. ball   d. cylinder   e. cylinder   f. cylinder   g. ball   h. cylinder

5. cylinder, ball

6. box, cube

7. Answers will vary. Please check the student's work.

8. Answers will vary. Please check the student's work.

9. a. ball   b. cylinder   c. box   d. cylinder

Review Chapter 6, p. 86

1.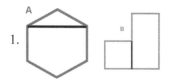

c: Answers may vary. One possibility:

2.

3.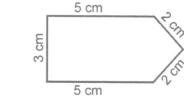

It is a quadrilateral (or, to be more precise, a parallelogram).

4. 5 corners.

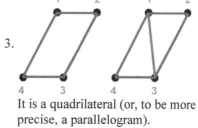

Chapter 7: Adding and Subtracting Within 0-100

Refresh Your Memory, p. 90

1.

7	8	9	10
5 and 2	1 and 7	4 and 5	3 and 7
6 and 1	4 and 4	8 and 1	9 and 1
1 and 6	5 and 3	2 and 7	6 and 4
2 and 5	7 and 1	1 and 8	5 and 5
4 and 3	2 and 6	3 and 6	2 and 8

2.

a.	b.	c.	d.
$7 - 1 = 6$	$8 - 3 = 5$	$9 - 2 = 7$	$10 - 3 = 7$
$7 - 5 = 2$	$8 - 6 = 2$	$9 - 3 = 6$	$10 - 8 = 2$
$7 - 3 = 4$	$8 - 2 = 6$	$9 - 5 = 4$	$10 - 5 = 5$
$7 - 6 = 1$	$8 - 7 = 1$	$9 - 7 = 2$	$10 - 6 = 4$

3.

a.	b.
$4 + 5 = 9$	$3 + 4 = 7$
$5 + 4 = 9$	$4 + 3 = 7$
$9 - 5 = 4$	$7 - 3 = 4$
$9 - 4 = 5$	$7 - 4 = 3$

4. a. 10, 9, 6   b. 9, 8, 3   c. 6, 1, 2

5. a. $2 \rightarrow 4 \rightarrow 6 \rightarrow 5 \rightarrow 4 \rightarrow 9 \rightarrow 6 \rightarrow 10 \rightarrow 3 \rightarrow 5 \rightarrow 1$
   b. $7 \rightarrow 10 \rightarrow 11 \rightarrow 9 \rightarrow 3 \rightarrow 6 \rightarrow 4 \rightarrow 10 \rightarrow 5 \rightarrow 11 \rightarrow 10$

6. a. 10, 20, 30, 40, 50, 60, 70, 80, 90
   b. 24, 34, 44, 54, 64, 74, 84, 94, 104
   c. 18, 28, 38, 48, 58, 68, 78, 88, 98

7. a. 26, 48, 52   b. 4, 8, 6   c. 20, 42, 77   d. 40, 56, 72

Adding Without Carrying, p. 92

1.

a. $31 + 3 = 34$	b. $32 + 6 = 38$
c. $43 + 3 = 46$	d. $47 + 2 = 49$

2.

a. $5 + 2 = 7$	b. $4 + 5 = 9$	c. $3 + 6 = 9$
$35 + 2 = 37$	$64 + 5 = 69$	$93 + 6 = 99$

3.

a. $52 + 7 = 59$	b. $33 + 1 = 34$	c. $11 + 5 = 16$
$\underline{2} + \underline{7} = 9$	$3 + 1 = 4$	$1 + 5 = 6$

4.

a. 35 + 3	b. 12 + 6	c. 57 + 1	d. 64 + 3
tens ones	tens ones	tens ones	tens ones
3　5	1　2	5　7	6　4
+　↓　3	+　↓　6	+　↓　1	+　↓　3
3　8	1　8	5　8	6　7

5.

a. 26 + 3	b. 72 + 4	c. 65 + 4	d. 81 + 4
tens ones	tens ones	tens ones	tens ones
2　6	7　2	6　5	8　1
+　↓　3	+　↓　4	+　↓　4	+　↓　4
2　9	7　6	6　9	8　5

6.

a.	b.	c.	d.
6 + 2 = 8	4 + 3 = 7	5 + 4 = 9	11 + 7 = 18
16 + 2 = 18	24 + 3 = 27	45 + 4 = 49	61 + 7 = 68
36 + 2 = 38	34 + 3 = 37	65 + 4 = 69	41 + 7 = 48

7.

a.	b.	c.
20 + 5 + 2 = 27	93 + 1 + 5 = 99	100 + 4 + 5 = 109
44 + 2 + 2 = 48	83 + 4 + 3 = 90	52 + 4 + 2 = 58

8.

a.	b.	c.
18 = 10 + 8	32 = 30 + 2	66 = 60 + 6
25 = 20 + 5	95 = 90 + 5	89 = 9 + 80
55 = 50 + 5	49 = 40 + 9	78 = 8 + 70

9.

a. 24 + 3 < 24 + 5	c. 17 + 2 < 19 + 2	e. 58 < 8 + 51
b. 83 + 5 = 85 + 3	d. 36 + 4 < 46 + 4	f. 66 = 5 + 61

Puzzle Corner. The one marked with ? is the comparison you cannot do without knowing the mystery number.
Why is that? Because the comparison depends on the value of the star. If ☆ is a large number, such as 100, then
☆ + ☆ > ☆ + 20. But if ☆ is a small number such as 2, then ☆ + ☆ < ☆ + 20.

☆ + 5　>　☆ + 4   ☆ − 5　<　☆ − 4   ☆ − 5　<　☆

☆ + 2　<　☆ + 7   ☆ − 5　>　☆ − 6   ☆ + ☆　?　☆ + 20

69

Subtracting Without Borrowing, p. 95

1. a. 6, 26   b. 1, 11   c. 0, 60   d. 0, 50   e. 1, 41   f. 3, 93

2. a. 52, 2   b. 74, 6 − 2 = 4   c. 84, 8 − 4 = 4

3. a. 31, 32, 33   b. 50, 52, 54   c. 46, 44, 42   d. 33, 32, 30

4. a. 71, 21   b. 45, 74   c. 53, 84   d. 12, 94

5. a. 5, 6, 7   b. 1, 1, 4   c. 4, 3, 7

6. a. She sold 28 pictures in total. 21 + 7 = 28
   b. She has two left. 28 + 2 = 30 or 30 − 28 = 2
   c. At 7:30. It took her three hours to paint the three pictures, and three hours later than 4:30 is 7:30.

7. a. 37 − 7 = 30   b. 46 − 6 = 40   e. 28 − 8 = 20
   d. 57 − 7 = 50   e. 85 − 5 = 80   f. 69 − 9 = 60

8.

a. 50 + 7 = 57	b. 86 + 2 = 88	c. 79 − 9 = 70
d. 25 − 5 = 20	e. 90 − 5 = 85	f. 42 = 40 + 2

9. a. 10, 15, 20, 25, 30, 35, 40, 45, 50
   b. 1, 6, 11, 16, 21, 26, 31, 36, 41
   c. 3, 8, 13, 18, 23, 28, 33, 38, 43

10.

a.	b.	c.
88 − 0 = 88	95 − 2 = 93	48 − 1 = 47
88 − 1 = 87	85 − 2 = 83	46 − 1 = 45
88 − 2 = 86	75 − 2 = 73	44 − 1 = 43
88 − 3 = 85	65 − 2 = 63	42 − 1 = 41
88 − 4 = 84	55 − 2 = 53	40 − 1 = 39
88 − 5 = 83	45 − 2 = 43	38 − 1 = 37
88 − 6 = 82	35 − 2 = 33	36 − 1 = 35
88 − 7 = 81	25 − 2 = 23	34 − 1 = 33

Adding or Subtracting Two-Digit Numbers, p. 98

1. a. 28   b. 6   c. 11   d. 15   e. 14   f. 22

2. a. 55   b. 51   c. 67   d. 48   e. 29   f. 66

3.

a.	b.	c.
35 + 20 = 55	40 + 17 = 57	33 − 20 = 13
76 + 30 = 106	30 + 33 = 63	78 − 50 = 28
22 + 50 = 72	56 − 20 = 36	99 − 40 = 59

4.

	a. tens	ones	b. tens	ones	c. tens	ones	d. tens	ones
	4	2	5	3	2	5	3	5
+	2	4		6	5	3		4
	6	6	5	9	7	8	3	9

5. a.

tens	ones
9	5
− 2	0
7	5

b.

tens	ones
5	8
− 2	6
3	2

c.

tens	ones
2	5
−	3
2	2

d.

tens	ones
7	9
− 6	4
1	5

6. a.

tens	ones
1	7
+ 2	1
3	8

b.

tens	ones
3	4
+ 1	4
4	8

c.

tens	ones
5	1
+	7
5	8

d.

tens	ones
5	1
+	7
5	8

7. a. 24   b. 13   c. 22   d. 12

8.

9. a. 50, 80, 80   b. 100, 60, 30   c. 20, 20, 50

10.

a. 57 − 21   b. 74 − 14   c. 59 − 7   d. 99 − 58

tens	ones
5	7
− 2	1
3	6

tens	ones
7	4
− 1	4
6	0

tens	ones
5	9
−	7
5	2

tens	ones
9	9
− 5	8
4	1

11.

a. Seventeen fish were put into the freezer.	−	2 8 1 1 1 7
b. You have to pay $56.	+	2 2 3 4 5 6
c. Mom is 27 years older than John.	−	3 8 1 1 2 7
d. Matt has 28 colored pencils now.	+	2 2 6 2 8

Review: What numbers make 10?	$1 + \underline{9} = 10$ $7 + \underline{3} = 10$ $4 + \underline{6} = 10$	$8 + \underline{2} = 10$ $5 + \underline{5} = 10$ $9 + \underline{1} = 10$	$3 + \underline{7} = 10$ $6 + \underline{4} = 10$ $2 + \underline{8} = 10$

1. a. $33 + 7 = 40$ b. $43 + 7 = 50$ c. $27 + 3 = 30$ d. $36 + 4 = 40$ e. $62 + 8 = 70$ f. $54 + 6 = 60$

2.

a. $\underline{10}$, 13, $\underline{20}$	b. $\underline{50}$, 57, $\underline{60}$	c. $\underline{40}$, 46, $\underline{50}$
d. $\underline{80}$, 81, $\underline{90}$	e. $\underline{70}$, 78, $\underline{80}$	f. $\underline{90}$, 94, $\underline{100}$

3. a. $56 + 4 = 60$ b. $35 + 5 = 40$ c. $49 + 1 = 50$

4.

a. $3 + \underline{7} = 10$ $23 + \underline{7} = 30$	b. $4 + \underline{6} = 10$ $44 + \underline{6} = \underline{50}$	c. $7 + \underline{3} = 10$ $17 + \underline{3} = \underline{20}$

5. a. 7 b. 9 c. 6 d. 2 e. 6 f. 4 g. 3 h. 5 i. 8 j. 2 k. 9 l. 1

6.

a. $36 + 4 = 40$ $40 - 4 = 36$	b. $57 + 3 = 60$ $60 - 3 = 57$	c. $83 + 7 = 90$ $90 - 7 = 83$
d. $66 + 4 = 70$ $70 - 4 = 66$		e. $95 + 5 = 100$ $100 - 5 = 95$

7. a. $30 + 7 + 3 = 40$.   Jeanine needs three more dollars.
   b. $20 + 10 + 10 = 40$. Derek needs ten more dollars.
   c. $12 + 20 + 8 = 40$. Muhammad needs eight more dollars.

Puzzle corner. Answers will vary because there are many possible solutions. These are just two example solutions.

100	−	10	−	50	= 40
−		+		+	
30	+	30	+	30	= 90
‖		‖		‖	
70		40		80	

100	−	30	−	30	= 40
−		+		+	
30	+	10	+	50	= 90
‖		‖		‖	
70		40		80	

1. a. 14 b. 17 c. 18 d. 13

2.

a. 7 + 8 = 15	b. 8 + 8 = 16
c. 6 + 5 = 11	d. 9 + 4 = 13
e. 8 + 5 = 13	f. 8 + 9 = 17
g. 7 + 7 = 14	h. 9 + 9 = 18

3. a. 13 + 9 = 22   b. 15 + 8 = 23   c. 17 + 7 = 24
   d. 24 + 7 = 31   e. 25 + 6 = 31   f. 37 + 9 = 46
   g. 36 + 6 = 42   h. 48 + 4 = 52   i. 58 + 5 = 63

4.

a. 28 + 8
      / \\
28 + _2_ + _6_
   30 + _6_ = _36_
	b. 47 + 5
	      / \\
	47 + _3_ + _2_
	   50 + _2_ = _52_
		c. 79 + 9
		      / \\
		79 + _1_ + _8_
		   80 + _8_ = _88_

d. 39 + 3
      / \\
39 + _1_ + _2_
   _40_ + _2_ = _42_
	e. 27 + 5
	      / \\
	27 + _3_ + _2_
	   _30_ + _2_ = _32_
		f. 38 + 7
		      / \\
		38 + _2_ + _5_
		   _40_ + _5_ = _45_

5. a. 40   b. 42   c. 64   d. 35   e. 62   f. 61

6. a. 39   b. 40   c. 52   d. 38   e. 59   f. 62

7. a.

		Count
Dad	卌卌 卌 IIII	19
Mom	卌 卌 卌 卌 卌 III	28
Mary	卌 卌 II	12
Mark	卌 卌 卌 卌 卌	25
Angie	卌 卌 卌 卌 卌 卌卌 I	36

c. Dad saw seven more birds than Mary.

d. Angie saw 11 more birds than Mark.

b.

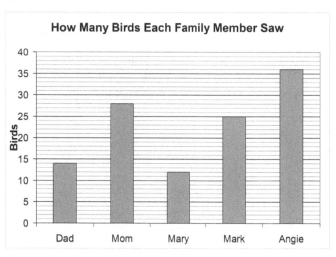

Subtracting from Whole Tens, p. 109

1. a. 36, 34, 33, 32
   b. 25, 26, 21, 24
   c. 48, 42, 47, 44
   d. 53, 51, 59, 56

2. a. $70 - 10 - 10 - 10 = 40$ b. $90 - 20 - 20 - 20 = 30$

3. $10 \rightarrow 60 \rightarrow 100 \rightarrow 80 \rightarrow 20 \rightarrow 40 \rightarrow 10 \rightarrow 30 \rightarrow 90 \rightarrow 60$

4. a. 64, 65, 68   b. 42, 43, 44   c. 39, 38, 37   d. 95, 93, 91

5.

a. $10 - 2 = 8$ $10 + 2 = 12$	b. $60 - 5 = 55$ $60 + 5 = 65$	c. $25 - 4 = 21$ $25 + 4 = 29$

6. a. $20 - 16 = 4$ pencils

   b. $17 - 7 = 10$ bushes

   c. $20 - 13 = 7$. Julie has 7 more stones than Carmen.
      $18 - 13 = 5$. Jane has 5 more stones than Carmen.
      $13 + 7 = 20$. Carmen needs seven more stones.

Add Using "Just One More", p. 111

1.

a. $8 + \underline{2} = 10$ $8 + \underline{3} = 11$	b. $4 + \underline{6} = 10$ $4 + 7 = 11$	c. $\underline{7} + 3 = 10$ $8 + 3 = 11$
d. $\underline{1} + 9 = 10$ $2 + 9 = 11$	e. $5 + \underline{5} = 10$ $5 + 6 = 11$	f. $\underline{4} + 4 = 8$ $5 + 4 = 9$

2. a. 3, 4   b. 2, 3   c. 4, 5   d. 6   e. 2   f. 8

3.

a. $7 + 2 = 9$ $3 + 8 = 11$ $5 + 5 = 10$	b. $5 + 6 = 11$ $3 + 4 = 7$ $6 + 4 = 10$	c. $4 + 6 = 10$ $2 + 8 = 10$ $7 + 4 = 11$	d. $2 + 9 = 11$ $5 + 4 = 9$ $3 + 7 = 10$

4. a. 13   b. 14   c. 17   d. 16   e. 11   f. 19   g. 15   h. 18   i. 11   j. 17   k. 13   l. 15

5. a. Joe gave away $3 + 2 + 5 = 10$ balloons. Joe still has $12 - 10 = 2$ balloons.
   b. Marsha found $7 + 6 = 13$ uniforms.
   c. She had to wash $13 - 3 = 10$ uniforms.
   d. There were $8 - 5 = 3$ more girls.
   e. There were just enough uniforms for everyone.

1.

a. 9 + 6 10 + 5 = 15	b. 9 + 4 10 + 3 = 13
c. 9 + 3 10 + 2 = 12	d. 9 + 5 10 + 4 = 14

2.

a. 9 + 8	b. 9 + 7	c. 9 + 9
/ \ 9 + 1 + 7 10 + 7 = 17	/ \ 9 + 1 + 6 10 + 6 = 16	/ \ 9 + 1 + 8 10 + 8 = 18

3.

a. 8 + 6 10 + 4 = 14	b. 8 + 7 10 + 5 = 15
c. 8 + 3 10 + 1 = 11	d. 8 + 4 10 + 2 = 12

4.

a. 8 + 8	b. 8 + 5	c. 8 + 7
/ \ 8 + 2 + 6 10 + 6 = 16	/ \ 8 + 2 + 3 10 + 3 = 13	/ \ 8 + 2 + 5 10 + 5 = 15

5. a. Not correct: 6 + 6 = 12.   b. Correct   c. Correct   d. Not correct: 9 + 7 = 16.

6. a. There are 9 − 2 − 1 = 6 apples left.
   b. Jeremy has 9 + 6 = 15 apples.
   c. Jeremy picked 9 − 7 = 2 more flowers. Altogether, they have 9 + 7 = 16 flowers.
   d. The line of cars was 5 + 5 + 4 = 14 cm long.

7.

a. 7 + 3 + 5 = 15	b. 9 + 1 + 2 = 12	c. 7 + 3 + 5 = 15
d. 6 + 4 + 6 = 16	e. 8 + 2 + 4 = 14	f. 5 + 5 + 8 = 18

8. a. 14   b. 15   c. 13   d. 12   e. 15   f. 18   g. 14   h. 16   i. 11

Puzzle corner: a. 8   b. 6   c. 4

Adding within 20, p. 116

a. 5 + 5 = 10	b. 6 + 6 = 12	c. 7 + 7 = 14
<u>5</u> + <u>6</u> = 11 and <u>6</u> + <u>5</u> = 11	6 + 7 = 13 and 7 + 6 = 13	7 + 8 = 15 and 8 + 7 = 15
d. 8 + 8 = 16	e. 9 + 9 = 18	f. 10 + 10 = 20
8 + 9 = 17 and 9 + 8 = 17	9 + 10 = 19 and 10 + 9 = 19	10 + 11 = 21 and 11 + 10 = 21

2. a. 10 + 7 = 17   b. 2 + 10 = 12   c. 10 + 4 = 14   d. 5 + 10 = 15

3.

a. 1 + 9 = 10	b. 3 + 7 = 10	c. 8 + 2 = 10
1 + 10 = 11 or 2 + 9 = 11	3 + 8 = 11 or 4 + 7 = 11	8 + 3 = 11 or 9 + 2 = 11
d. 6 + 4 = 10	e. 5 + 5 = 10	f. 7 + 3 = 10
6 + 5 = 11 or 7 + 4 = 11	5 + 6 = 11 or 6 + 5 = 11	7 + 4 = 11 or 8 + 3 = 11

4. a. 7 + 7 = 14   Doubles chart
   b. 9 + 7 = 16   Trick with nine
   c. 8 + 3 = 11   Trick with eight
   d. 6 + 7 = 13   Just one more than a double
   e. 5 + 6 = 11   Just one more than a sum with 10
   f. 5 + 8 = 13   Trick with eight
   g. 8 + 8 = 16   Doubles chart
   h. 4 + 9 = 13   Trick with nine

5. a. 9 + 5 − 2 = 12   Maria has twelve dollars.
   b. 9 + 3 = 12; 8 + 3 = 11   Ashley has one more shirt.
   c. 10 − 6 − 1 = 3   Emily has three dollars.
   d. 8 − 2 + 4 = 10   They have 10 tennis balls.

6.

a. 8 + 2 = 10	b. 5 + 3 = 8	c. 9 + 2 = 11	d. 7 + 3 = 10
8 + 4 = 12	5 + 5 = 10	9 + 4 = 13	7 + 5 = 12
8 + 6 = 14	5 + 7 = 12	9 + 6 = 15	7 + 7 = 14
8 + 8 = 16	5 + 9 = 14	9 + 8 = 17	7 + 9 = 16

7.

8. a. 18, 28, 38, 48, 58, 68, 78, 88, 98
   b. 27, 37, 47, 57, 67, 77, 87, 97, 107

9.

a. 6 + 6 = 12	b. 8 + 8 = 16	c. 6 + 5 = 11
6 + 7 = 13	9 + 7 = 16	7 + 4 = 11

10. 9 + 6 = 15 cm   Check the student's lines.

Adding within 20, cont.

11.

a.	b.	c.	d.
8 + 8 = 16	7 + 8 = 15	7 + 7 = 14	5 + 8 = 13
2 + 9 = 11	9 + 6 = 15	9 + 8 = 17	3 + 9 = 12
7 + 5 = 12	6 + 5 = 11	7 + 4 = 11	7 + 6 = 13
e.	**f.**	**g.**	**h.**
9 + 4 = 13	8 + 6 = 14	9 + 2 = 11	6 + 9 = 15
4 + 8 = 12	6 + 6 = 12	8 + 5 = 13	8 + 7 = 15
6 + 7 = 13	5 + 9 = 14	5 + 7 = 12	8 + 4 = 12
i.	**j.**	**k.**	**l.**
9 + 3 = 12	4 + 9 = 13	9 + 9 = 18	8 + 9 = 17
4 + 7 = 11	7 + 7 = 14	6 + 8 = 14	5 + 6 = 11
9 + 5 = 14	3 + 8 = 11	6 + 6 = 12	8 + 3 = 11

Puzzle corner:
Answers may
vary. Check
the student's
work.

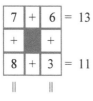

Subtract to 10, p. 120

1. b. 14 − 4 = 10   c. 16 − 6 = 10   d. 15 − 5 = 10

2. a. 13 − 3 = 10   b. 17 − 7 = 10   c. 19 − 9 = 10

3.

a. 14 − 7	b. 15 − 8	c. 16 − 8
14 − 4 − 3 10 − 3 = 7	15 − 5 − 3 10 − 3 = 7	16 − 6 − 2 10 − 2 = 8
d. 13 − 6	**e. 12 − 6**	**f. 13 − 4**
13 − 3 − 3 10 − 3 = 7	12 − 2 − 4 10 − 4 = 6	13 − 3 − 1 10 − 1 = 9

4.

a. 12 − 6	b. 15 − 9	c. 13 − 8
12 − 2 − 4 = 6	15 − 5 − 4 = 6	13 − 3 − 5 = 5
d. 13 − 7	**e. 14 − 7**	**f. 12 − 4**
13 − 3 − 4 = 6	14 − 4 − 3 = 7	12 − 2 − 2 = 8

5. a. 7   b. 8   c. 7   d. 8   e. 7   f. 9   g. 8   h. 5

6. a. 13 − 8 = 5   Tom is five years older than Juan.
   b. 13 − 9 = 4   Tom is four years older than Alice.
   c. 15 − 10 = 5   Tom will still be five years older than Juan.

7.

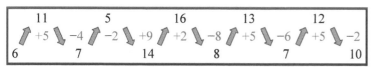

77

1.

a. 8 + 5 = 13 5 + 8 = 13 13 − 8 = 5 13 − 5 = 8	b. 9 + 7 = 16 7 + 9 = 16 16 − 9 = 7 16 − 7 = 9

2.

a. 8 + 4 = 12 12 − 8 = 4 12 − 4 = 8	b. 9 + 7 = 16 16 − 9 = 7 16 − 7 = 9	c. 7 + 6 = 13 13 − 7 = 6 13 − 6 = 7

3.

a. 11 − 3 = 8 3 + 8 = 11	b. 11 − 4 = 7 4 + 7 = 11	c. 12 − 3 = 9 3 + 9 = 12

4.

a. 14 − 8 = 6 8 + 6 = 14	b. 15 − 7 = 8 7 + 8 = 15	c. 17 − 8 = 9 8 + 9 = 17
d. 12 − 8 = 4 8 + 4 = 12	e. 16 − 7 = 9 7 + 9 = 16	f. 13 − 7 = 6 7 + 6 = 13
g. 13 − 8 = 5 8 + 5 = 13	h. 11 − 7 = 4 7 + 4 = 11	i. 14 − 9 = 5 9 + 5 = 14

5.

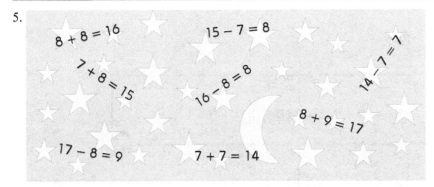

The matching additions and subtractions are:
8 + 8 = 16 and 16 − 8 = 8
7 + 8 = 15 and 15 − 7 = 8
8 + 9 = 17 and 17 − 8 = 9
7 + 7 = 14 and 14 − 7 = 7

6.

a. 12 − 8 = 4	b. 11 − 7 = 4	c. 13 − 9 = 4
d. 15 − 6 = 9	e. 18 − 9 = 9	f. 16 − 7 = 9

7. a. 15 − 6 = 9 Marsha has nine crayons.
    6 + 6 = 12 Susana has 12 crayons now.
    Susana has more crayons.
    Susana has 12 − 9 = 3 more crayons.
  b. 7 + 8 = 15. She has 15 stars in her drawing.
  c. 12 − 7 = 5. Matthew has $5.
  d. 6 + 8 = 14. Neither is right. The total is $14.

Using Addition to Subtract, cont.

8.

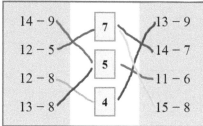
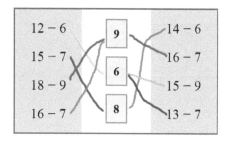

Some Mixed Review, p. 125

1. a. 9:30   b. 12:30   c. 6:00   d. 3:30

2.

Now it is:	a. 2:00	b. 8:00	c. 12:00	d. 7:30	e. 10:30
half-hour later	2:30	8:30	12:30	8:00	11:00

3.

a.	b.	c.
100 − 1 = 99	10 − 1 = 9	10 + 90 = 100
90 − 2 = 88	20 − 2 = 18	20 + 80 = 100
80 − 3 = 77	30 − 3 = 27	30 + 70 = 100
70 − 4 = 66	40 − 4 = 36	40 + 60 = 100
60 − 5 = 55	50 − 5 = 45	50 + 50 = 100
50 − 6 = 44	60 − 6 = 54	60 + 40 = 100
40 − 7 = 33	70 − 7 = 63	70 + 30 = 100
30 − 8 = 22	80 − 8 = 72	80 + 20 = 100

4.

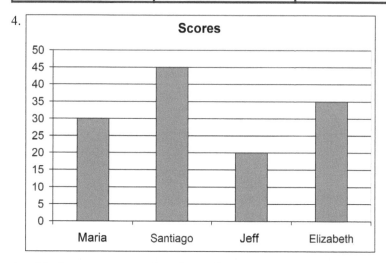

4. a. Maria got ten more points than Jeff.
   b. Santiago got ten more points than Elizabeth.

5. Answers will vary. Check the student's answers.

6. Answers will vary. Check the student's answers.

Some Mixed Review, cont.

7.

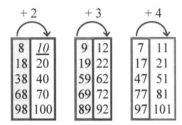

Puzzle corner:

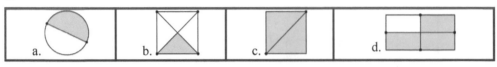

Pictographs, p. 128

1. a. <u>Jim</u> rode the most miles. Jim rode <u>70 miles</u>
   b. The boys that rode the least miles were Greg and <u>Ernest</u>. Greg rode 15 miles. Ernest rode 25 miles.
   c. Matthew rode <u>10 more miles</u> than Dan.
   d. Dan rode <u>25 more miles</u> than Greg.

2.

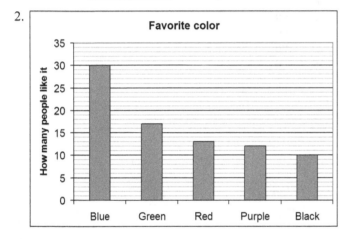

3. a.

	How many?	
oranges	15	
mangos	21	
bananas	24	

   b. 15 + 24 = 39. They picked a total of 39 oranges and bananas.
   c. 21 + 3 = 24 or 24 − 21 = 3. Or, you can note that there is half a fruit more in the pictograph for mangos than for bananas, which means 3 pieces of fruit. So, there were three more bananas than mangos.

4. Answers will vary. Check the student's work. Possible responses include:

   How many points did Mark get? How many points did Aaron get?
   How many more points did Aaron get than Jack?
   Who got the most points? Who got the least points?
   How many points did the winner get?
   How many points more did the winner get than Jack?
   How many points did Mark and Aaron get in total?

1. YOU FOUND ALL OF THEM!

2.

a. 31 + 45

	3	1
+	4	5
	7	6

b. 70 + 19

	7	0
+	1	9
	8	9

c. 26 + 73

	2	6
+	7	3
	9	9

d. 31 + 8

	3	1
+		8
	3	9

e. 77 − 22

	7	7
−	2	2
	5	5

f. 56 − 14

	5	6
−	1	4
	4	2

g. 99 − 45

	9	9
−	4	5
	5	4

h. 47 − 5

	4	7
−		5
	4	2

3. b. You can use the method of finding the double
and adding one for the problems below.

7 + 8 = 15   6 + 7 = 13
6 + 5 = 11   8 + 9 = 17

5 + 5 = 10
6 + 6 = 12
7 + 7 = 14
8 + 8 = 16
9 + 9 = 18

4. a. 9 + 9 = 18   Doubles chart OR Trick with nine
   b. 8 + 4 = 12   Trick with eight
   c. 9 + 5 = 14   Trick with nine
   d. 7 + 7 = 14   Doubles chart
   e. 7 + 8 = 15   "Just one more" than a double OR Trick with eight
   f. 6 + 5 = 11   "Just one more" than a sum of 10 OR "Just one more" than a double
   g. 3 + 9 = 12   Trick with nine
   h. 6 + 7 = 13   "Just one more" than a double

5.

a. 11 − 2 = 9 11 − 4 = 7 11 − 5 = 6 11 − 6 = 5	b. 12 − 4 = 8 12 − 5 = 7 12 − 3 = 9 12 − 6 = 6	c. 13 − 5 = 8 13 − 6 = 7 13 − 4 = 9 13 − 7 = 6
d. 14 − 5 = 9 14 − 8 = 6 14 − 7 = 7 14 − 6 = 8	e. 15 − 6 = 9 15 − 9 = 6 15 − 7 = 8 15 − 8 = 7	f. 16 − 8 = 8 16 − 9 = 7 16 − 7 = 9 16 − 6 = 10

6. a. Mariana read 15 books.
   b. Jose read 30 books.
   c. Janet read 20 − 10 = 10 more books than Jim.
   d. Jose read 30 − 20 = 10 more books than Janet.
   e. Answers will vary.

7. a. 20 − 2 − 5 = 13 birds are left.
   b. 5 + 1 + 3 = 9 books.
   c. 20 − 14 = 6 pages left.
   d. 12 − 4 = 8 Sam is eight years older.
   e. $11 + $5 = $16. Yes, I will have enough with a dollar left over.

8.

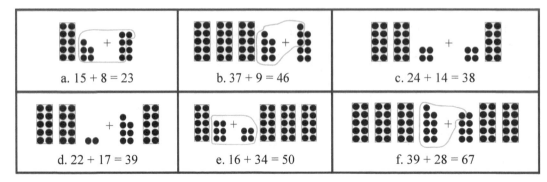

a. 15 + 8 = 23	b. 37 + 9 = 46	c. 24 + 14 = 38
d. 22 + 17 = 39	e. 16 + 34 = 50	f. 39 + 28 = 67

9.

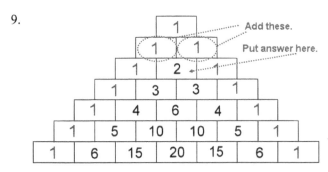

Add these.

Put answer here.

```
                1
            1       1
          1     2     1
        1    3     3    1
      1    4    6    4    1
    1    5   10   10    5   1
  1    6   15   20   15   6   1
```

Puzzle corner

a.

	4	5
−	2	3
	2	2

b.

	7	9
−	6	4
	1	5

c.

	3	6
−		4
	3	2

d.

	5	7
−	1	7
	4	0

e.

	6	7
−	1	5
	5	2

Chapter: 8 Coins

Counting Dimes, Nickels, and Cents, p. 136

1. a. 30¢   b. 23¢   c. 62¢   d. 73¢   e. 42¢   f. 37¢   g. 60¢

2. a. one dime, two pennies   b. four dimes
   c. two dimes, four pennies   d. three dimes, one penny

3. a. 15¢   b. 18¢   c. 25¢   d. 37¢   e. 35¢   f. 9¢

4. a. 15¢   b. 45¢   c. 30¢   d. 60¢   e. 25¢   f. 45¢   g. 6¢   h. 17¢   i. 40¢   j. 17¢   k. 29¢   l. 55¢

Counting Dimes, Nickels, and Cents 2, p. 139

1. a. 20¢   b. 47¢   c. 68¢   d. 48¢   e. 76¢   f. 52¢

2. a. 12¢ now   b. 21¢ now   c. 41¢ now   d. 21¢ now   e. 17¢ now   f. 24¢ now

3. a. 16¢ now   b. 26¢ now   c. 40¢ now   d. 33¢ now

4. a. two dimes, one nickel
   b. three dimes, one nickel, four pennies;
   c. one dime, four pennies
   d. three dimes, one nickel, three pennies;
   e. six dimes, three pennies
   f. one dime, one nickel, one penny;
   g. six dimes, one penny
   h. four dimes, one nickel;
    i. two dimes, one nickel, two pennies

5. a. 20¢, 21¢, 23¢, 25¢, 26¢
   b. 26¢, 29¢, 30¢, 25¢, 32¢
   c. 60¢, 63¢, 60¢, 76¢, 88¢

Quarters, p. 141

1. a. 50¢   b. 75¢   c. 100¢   d. 45¢   e. 55¢   f. 70¢   g. 95¢   h. 80¢   i. 85¢

2. a. 30¢   b. 80¢   c. 40¢

3. a. 31¢   b. 44¢   c. 75¢   d. 35¢   e. 51¢   f. 55¢   g. 40¢   h. 77¢   i. 98¢   j. 53¢   k. 91¢   l. 78¢

4. a. 45¢   b. 40¢   c. 21¢   d. 87¢

5. a. Left 15¢   b. Left 6¢   c. Left 15¢   d. Left 23¢   e. Left 13¢
   f. Left 63¢   g. Left 31¢   h. Left 12¢   i. Left 43¢

Practicing with Money, p. 144

1. a. one quarter, four pennies
   b. one quarter, two dimes, one penny
   c. two quarters, one dime, two pennies
   d. one quarter, two dimes, three pennies
   e. three quarters, one dime (or two nickels), one penny
   f. three quarters, one dime, one nickel, one penny

2. a. one quarter, one nickel
   b. one quarter, one nickel, two pennies
   c. one quarter, one dime
   d. one quarter, two dimes
   e. one quarter, one dime, one nickel, one penny
   f. one quarter, two dimes, three pennies

Practicing with Money, cont.

3. a. two quarters
   b. two quarters, three pennies
   c. two quarters, one nickel, three pennies
   d. two quarters, one dime
   e. two quarters, one dime, one nickel, one penny
   f. two quarters, two dimes, two pennies

Review—Coins, p. 146

1. a. 11¢   b. 27¢   c. 60¢   d. 32¢   e. 46¢   f. 77¢

2. a. two quarters, two pennies
   b. two dimes, one nickel, two pennies <u>or</u> one quarter and two pennies
   c. three quarters, one penny
   d. three quarters, one dime
   e. three quarters, four pennies
   f. three dimes, four pennies

3. a. 56¢   b. 51¢

Test Answer Keys

Math Mammoth Grade 1 Tests Answer Key

Chapter 1 Test

Grading

My suggestion for grading the chapter 1 test is below. The total is 28 points. Divide the student's score by the total of 28 to get a decimal number, and change that decimal to percent to get the student's percentage score.

Question	Max. points	Student score
1	6 points	
2	8 points	
3	6 points	

Question	Max. points	Student score
4	4 points	
5	4 points	
Total	28 points	

1. a. 6   b. 9   c. 8   d. 10   e. 9   f. 9

2. a. <   b. <   c. <   d. =   e. >   f. >   g. <   h. =

3. a. $4 + 6 = 10$ b. $3 + 4 = 7$ c. $5 + 5 = 10$

4. 5   b. 3   c. 6   d. 2

5. a. They have 10 stuffed animals. $7 + 3 = 10$ b. Six pairs are not in the wash. $2 + 6 = 8$ (or $8 - 2 = 6$).

Chapter 2 Test

Grading

My suggestion for grading the chapter 2 test is below. The total is 30 points. Divide the student's score by the total of 30 to get a decimal number, and change that decimal to percent to get the student's percentage score.

Question	Max. points	Student score
1	4 points	
2	4 points	
3	6 points	

Question	Max. points	Student score
4	16 points	
Total	30 points	

1. $2 + 6 = 8$; $6 + 2 = 8$; $8 - 2 = 6$; $8 - 6 = 2$

2. a. $9 - 5 = 4$ or $9 - 4 = 5$ b. $6 + 4 = 10$; $10 - 6 = 4$

3. a. There are 6 cats.   b.   c. There are 5 more robins than sparrows.

4. a. 2, 7, 5, 3   b. 6, 2, 7, 6   c. 9, 6, 2, 6   d. 1, 1, 0, 3

Chapter 3 Test

Grading

My suggestion for grading the chapter 3 test is below. The total is 27 points. Divide the student's score by the total of 27 to get a decimal number, and change that decimal to percent to get the student's percentage score.

Question	Max. points	Student score
1	4 points	
2	5 points	
3	5 points	
4	4 points	

Question	Max. points	Student score
5	6 points	
6	3 points	
Total	27 points	

1. a. sixteen   b. seventy-eight   c. fifty-one   d. ninety

2.

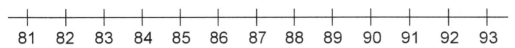

3. a. $86 = 80 + 6$ b. $52 = 50 + 2$, $32 = 30 + 2$ c. $97 = 90 + 7$, $19 = 10 + 9$

4. a. 29   b. 75   c. 82   d. 91

5. a. $57 < 71 < 75$ b. $69 < 96 < 98$ c. $49 < 81 < 84$

6. a. $=$ b. $<$ c. $>$

Chapter 4 Test

Grading

My suggestion for grading the chapter 4 test is below. The total is 32 points. Divide the student's score by the total of 32 to get a decimal number, and change that decimal to percent to get the student's percentage score.

Question	Max. points	Student score
1	14 points	
2	6 points	

Question	Max. points	Student score
3	12 points	
Total	32 points	

1. a. 6, 5, 7, 4   b. 7, 6, 4, 8   c. 8, 7, 4, 9   d. 3, 1, 6, 5   e. 1, 0, 2, 5   f. 6, 1, 3, 2   g. 4, 1, 1, 3

2. a. Liz now has $3 + 5 = 8$. So Liz has $8 - 7 = 1$ more.
   b. $2 + 4 = 6$ and $6 - 3 = 3$. So Dan now has 3 boxes of nails.

3.

a. 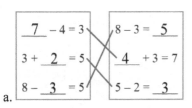   b.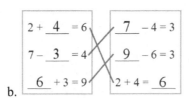

Chapter 5 Test

Grading

My suggestion for grading the chapter 5 test is below. The total is 34 points. Divide the student's score by the total of 34 to get a decimal number, and change that decimal to percent to get the student's percentage score.

Question	Max. points	Student score
1	8 points	
2	12 points	
3	10 points	

Question	Max. points	Student score
4	4 points	
Total	34 points	

1. a. 1 o'clock   b. half past 5   c. 7 o'clock   d. half past 12

2. a. half past 3 or 3:30   b. 9 o'clock or 9:00   c. half past 2 or 2:30   d. half past 11 or 11:30

3.

Now it is:	a. 6:00	b. 9:30	c. 10:00	d. 4:30	e. 12:30
a half-hour later, it is:	6:30	10:00	10:30	5:00	1:00
an hour later, it is:	7:00	10:30	11:00	5:30	1:30

4. a. AM   b. PM   c. AM   d. PM

Chapter 6 Test

Grading

My suggestion for grading the chapter 6 test is below. The total is 9 points. Divide the student's score by the total of 9 to get a decimal number, and change that decimal to percent to get the student's percentage score.

Question	Max. points	Student score
1	2 points	
2	2 points	
3	3 points	

Question	Max. points	Student score
4	2 points	
Total	9 points	

1. a. rectangle   b. triangle

2.

3.

The bottom side is 2 inches, the other two are 3 inches.
(image not to scale) It is a triangle.

4. a. 4 inches.

b. 12 centimeters.

Grading

My suggestion for grading the chapter 7 test is below. The total is 33 points. Divide the student's score by the total of 33 to get a decimal number, and change that decimal to percent to get the student's percentage score.

Question	Max. points	Student score
1	3 points	
2	6 points	
3	2 points	

Question	Max. points	Student score
4	8 points	
5	6 points	
Total	33 points	

1. a. 26, 46   b. 70, 86   c. 70, 10

2.

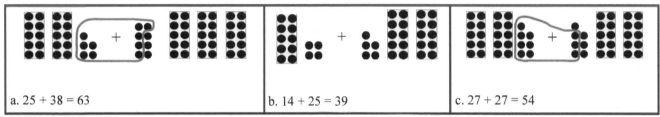

a. 25 + 38 = 63 | b. 14 + 25 = 39 | c. 27 + 27 = 54

3. a. 18   b. 13   c. 13   d. 15

4. a. 20 + 57   b. 78 − 44   c. 45 + 13   d. 87 − 20

	2	0
+	5	7
	7	7

	7	8
−	4	4
	3	4

	4	5
+	1	3
	5	8

	8	7
−	2	0
	6	7

5. Jake has more money (he has now $16). Jake has 4 dollars more than Jim.

Chapter 8 Test

Grading

My suggestion for grading the chapter 8 test is below. The total is 13 points. Divide the student's score by the total of 13 to get a decimal number, and change that decimal to percent to get the student's percentage score.

Question	Max. points	Student score
1	6 points	
2	3 points	

Question	Max. points	Student score
3	4 points	
Total	13 points	

1. a. 11¢   b. 34¢   c. 39¢   d. 42¢   e. 58¢   f. 102¢

2. Answers may vary. For example:
   a. Two quarters, one dime, and three pennies.   b. Three dimes, one nickel, and three pennies.
   c. Two quarters, one dime, one nickel, and four pennies.

3. a. 6¢ left.   b. 23¢ left.

End-of-the-Year Test Grade 1 Answer Key

Instructions to the teacher:

My suggestion for grading is below. The total is 104 points. A score of 83 points is 80%. A score of 73 points is 70%.

Question	Max. points	Student score
Basic Addition and Subtraction Facts within 0-10		
1	8 points	
2	8 points	
3	4 points	
4	8 points	
	subtotal	/ 28
Place Value and Two-Digit Numbers		
5	6 points	
6	4 points	
7	3 points	
	subtotal	/ 13
Adding and Subtracting Two-Digit Numbers		
8	6 points	
9	6 points	
10	4 points	
11	3 points	
	subtotal	/ 19

Question	Max. points	Student score
Basic Word Problems		
12	2 points	
13	2 points	
14	2 points	
15	2 points	
16	2 points	
17	6 points	
18	6 points	
	subtotal	/ 22
Clock		
19	6 points	
20	4 points	
	subtotal	/ 10
Geometry and Measuring		
21	2 points	
22	5 points	
	subtotal	/ 7
Money		
23	3 points	
24	2 points	
	subtotal	/ 5
	TOTAL	/ 104

1. a. 5, 8, 7, 9   b. 10, 9, 9, 8   c. 8, 10, 7, 8   d. 10, 6, 10, 8

2. a. 5, 2, 4, 1   b. 2, 3, 3, 3   c. 4, 5, 1, 3   d. 7, 1, 2, 2

3. a. $2 + 7 = 9$; $7 + 2 = 9$; $9 - 2 = 7$; $9 - 7 = 2$

4. a. 5, 5   b. 7, 8   c. 2, 6   d. 5, 4

5. a. 27, 65   b. 50, 9   c. 0, 90

6. a. $16 < 26 < 61$ b. $14 < 51 < 54$

7. a. <   b. <   c. =

8. a. 88, 45   b. 76, 18   c. 79, 59

9. a. 50, 14   b. 52, 60   c. 26, 48

10. a. 49   b. 25   c. 96   d. 36

11. The child can circle some of the dots to make a ten in (a) and (b). That makes it easier to see the total.
    a. 53   b. 50   c. 49

12. $14 - 8 = 6$ or $14 - 6 = 8$

13. 20 more

14. Mark's cars 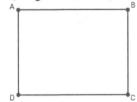 Henry's cars

15. $10 - 6 = 4$ girls

16. The books cost $10 + $5 = $15. Andy has left: $20 − $15 = $5.

17. a. 8 spaces   b. 24 cars   c. 6 spaces

18. a. Isabelle has now 60 marbles.   b. Her sister has 65 marbles.
    c. Her sister has more; five more marbles.

19. a. 11 o'clock, 11:00   b. half past one, 1:30   c. half past 8, 8:30

20.

Now it is:	a. 5:30	b. 12:00
a half-hour later, it is:	6:00	12:30
an hour later, it is:	6:30	1:00

21. a.

    b.

22. a. (image not to scale)

    b. a rectangle
    c. Side AB: _8_ cm   Side BC: _6_ cm

22. d. triangles: (image not to scale)

23. a. 18¢   b. 42¢   c. 85¢

24. 69¢

Cumulative Reviews
Answer Keys

Grade 1 Cumulative Reviews Answer Key

Cumulative Review: Chapters 1 - 2

1. a. 8

   b. 5

   c. 4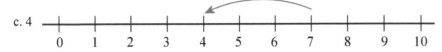

2. a. $4 + 6 = 10$ $10 - 4 = 6$
   b. $5 + 3 = 8$ $8 - 5 = 3$

3. a. 7   b. 10   c. 4   d. 9   e. 7

4. a. 7   b. 6   c. 7

5. a. =   b. <   c. <   d. =   e. =   f. >

6. a. 10, 8   b. 9, 8   c. 7, 5   d. 7, 5

7.

a. $6 + 1 + 3 = 10$	b. $1 + 4 + 3 = 8$

8.

a. Numbers: 9, 5, 4	b. Numbers: 10, 2, 8
$5 + 4 = 9$	$2 + 8 = 10$
$4 + 5 = 9$	$8 + 2 = 10$
$9 - 5 = 4$	$10 - 2 = 8$
$9 - 4 = 5$	$10 - 8 = 2$

9.

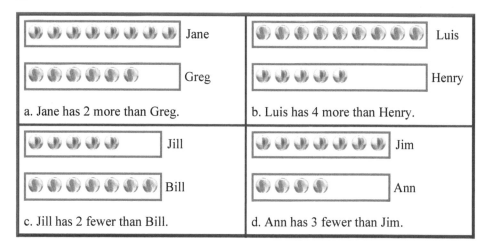

a. Jane has 2 more than Greg.

b. Luis has 4 more than Henry.

c. Jill has 2 fewer than Bill.

d. Ann has 3 fewer than Jim.

10. a. 3 + 5 − 1 = 7 children
    b. 3 + 7 = 10 marbles together; 7 − 3 = 4 more than Judy
    c. 10 − 7 = 3 blue trucks
    d. 9 − 6 = 3 more dollars
    e. 3 + 5 = 8 socks. Matt found 10 − 8 = 2 socks still missing

1. a. $22 = 20 + 2$　　b. $64 = 60 + 4$　　c. $95 = 90 + 5$

2. a. $<$　b. $=$　c. $>$　d. $<$　e. $=$　f. $=$

3.

a. $3 + 7 = 10$	b. $6 + 3 = 9$
$7 + 3 = 10$	$3 + 6 = 9$
$10 - 7 = 3$	$9 - 3 = 6$
$10 - 3 = 7$	$9 - 6 = 3$

4. a. 4, 14, 24, 34, 44, 54, 64, 74
   b. 38, 48, 58, 68, 78, 88, 98, 108

5. b. 50, 60 and 70 will be green if your blue and yellow blend together.

41	42	43	44	45	46	47	48	49	50
51	52	53	54	55	56	57	58	59	60
61	62	63	64	65	66	67	68	69	70

6. a. eleven　b. seventeen　c. fifteen　d. thirteen

7.

From	2	11	9	14	6	12	6	10
To	10	7	9	7	6	5	12	15
Difference	8	4	0	7	0	7	6	5

8. a. $>$　b. $>$　c. $>$　d. $<$　e. $<$　f. $>$

9. a. $10 - 2 - 2 = 6$. They need 6 more children.
   b. $10 + 20 + 10 = 40$ horses; $20 - 10 = 10$ more white than brown horses.

1.

2 + 3 < 6	1 + 6 > 6	4 + 3 < 8

2. a. 0 + 4 + 2 = 6   b. 7 + 1 + 1 = 9   c. 2 + 5 + 3 = 10

3.

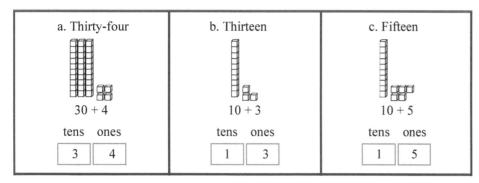

a. Thirty-four	b. Thirteen	c. Fifteen
30 + 4	10 + 3	10 + 5
tens ones	tens ones	tens ones
3 | 4	1 | 3	1 | 5

4.

a. 40 + _8_ = 48	b. 60 + _2_ = _62_	c. 50 + 5 = _55_

5. a. 17; 28; 39   b. 13; 65; 61   c. 22; 20; 97

6. 96, 97, 98 , 99 , 100 , 101 , 102 , 103 , 104 , 105 , 106, 107 , 108

7.

a. 9 ⌖ ||||	b. 11 ⌖ ⌖|
c. 27 ⌖ ⌖ ⌖ ⌖ ⌖||	d. 32 ⌖ ⌖ ⌖ ⌖ ⌖ ⌖||

8. a. Alice has 31.   b. Aaron has 12.   c. Maria has 17.   d. Now Alice has 21 and Aaron has 22.

9. a. 83   b. 72

Cumulative Review: Chapters 1 - 5

1. a. 5   b. 7   c. 9   d. 10

2. a. $5 - 2 = 3$ b. $9 - 4 = 5$ c. $7 - 3 = 4$ d. $10 - 8 = 2$

   e. $6 - 2 = 4$ f. $10 - 7 = 3$ g. $7 - 7 = 0$ h. $9 - 5 = 4$

3. two tens   twenty
   three tens   thirty
   eight tens   eighty
   five tens   fifty

4. a. $17 < 36 < 58$ b. $23 < 36 < 63$ c. $44 < 48 < 84$

5. a. 3   b. 6   c. 10   d. 6   e. 7   f. 0

6.

	a. two o'clock	b. ten o'clock	c. half-past six	d. half-past eight
1/2 hour later →	half past two	half past ten	seven o'clock	nine o'clock

7. a. When you woke up. It was 7 <u>AM</u>.	b. Jon plays in the afternoon at 3 <u>PM</u>.
c. Joe is asleep. It is dark. It is 1 <u>AM</u>.	d. It is time for lunch. It is 1 <u>PM</u>.

8. a. <   b. >   c. >   d. >   e. =   f. <

9. a. $10 - 4 = 6$ crayons in the bucket.   b. $10 + 4 = 14$ dollars. $14 + 6 = 20$ She needs six more dollars.

1. a. 0   b. 3   c. 5   d. 2

2.   ~~4 — 6~~   2 – 0   ~~6 — 9~~   8 – 4

3. a. twenty-nine   b. ninety-one   c. fifteen   d. fifty-seven

4.

Now it is:	a. 1:00	b. 11:30	c. 9:00	d. 6:30	e. 4:00
a half-hour later, it is:	1:30	12:00	9:30	7:00	4:30

5. Check the student's work. Answers will vary.

6.

7.

| $9 - 5 = 4$ |
| $3 + 1 = 4$ |
| $10 - 6 = 4$ |
| $2 + 2 = 4$ |

| $3 + 4 = 7$ |
| $8 - 1 = 7$ |
| $0 + 7 = 7$ |
| $9 - 2 = 7$ |

| $10 - 2 = 8$ |
| $3 + 5 = 8$ |
| $9 - 1 = 8$ |
| $2 + 6 = 8$ |

| $9 - 3 = 6$ |
| $3 + 3 = 6$ |
| $10 - 4 = 6$ |
| $2 + 4 = 6$ |

| $3 + 2 = 5$ |
| $10 - 5 = 5$ |
| $0 + 5 = 5$ |
| $9 - 4 = 5$ |

| $10 - 7 = 3$ |
| $3 + 0 = 3$ |
| $7 - 4 = 3$ |
| $2 + 1 = 3$ |

8. a. 30 < 38   b. 87 > 85   c. 69 < 96   d. 58 > 56

   e. 60 > 48   f. 43 < 95   g. 49 < 94   h. 22 < 32

1. a. $2 + 2 = 4$ $5 + 4 = 9$	b. $2 + 5 = 7$ $0 + 5 = 5$	c. $6 - 0 = 6$ $8 - 4 = 4$	d. $3 - 1 = 2$ $10 - 3 = 7$

2. a. 33   b. 86

3. 16, 26, 36, 46, 56, 66, 76, 86

4. a. $77 for both   b. $99 for both

5.

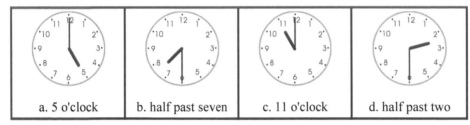

a. 5 o'clock	b. half past seven	c. 11 o'clock	d. half past two

6. a. a square or a rhombus   b. a triangle

7.

a. $9 + 8 = 17$ $8 + 8 = 16$	b. $9 + 3 = 12$ $8 + 4 = 12$	c. $9 + 5 = 14$ $7 + 8 = 15$

8.

a. $25 + 10 = 35$ $60 + 20 = 80$	b. $90 - 30 = 60$ $100 - 70 = 30$	c. $92 - 10 = 82$ $64 - 10 = 54$

9.

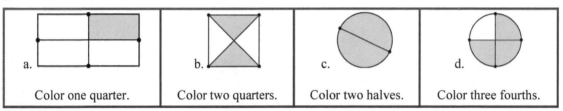

a. Color one quarter.	b. Color two quarters.	c. Color two halves.	d. Color three fourths.

10. They are equal.

11. a. cylinder b. cube   c. cylinder   d. box

Cumulative Review: Chapters 1 - 8

1.

a.	b.	c.	d.
10 − 1 = 9	8 + 9 = 17	25 − 10 = 15	52 + 7 = 59
8 − 2 = 6	7 + 8 = 15	38 − 10 = 28	35 + 3 = 38
7 − 3 = 4	5 + 6 = 11	100 − 10 = 90	26 + 2 = 28

2.

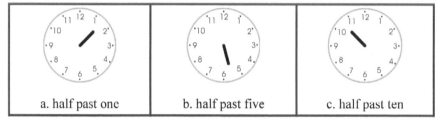

a. half past one	b. half past five	c. half past ten

3. Check the student's work.

4.

a. 15 − 7	b. 14 − 9	c. 16 − 8
/ \	/ \	/ \
15 − 5 − 2	14 − 4 − 5	16 − 6 − 2
= 8	= 5	= 8

5. a. 55   b. 85   c. 63   d. 42

6.

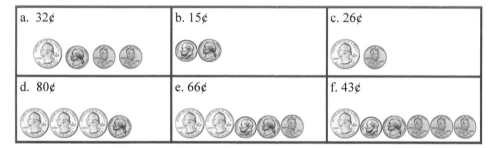

a. 32¢	b. 15¢	c. 26¢
d. 80¢	e. 66¢	f. 43¢

8. a. 26¢   b. 27¢